The Call of the Crawfish Frog

The Call of the Crawfish Frog

Michael J. Lannoo
and
Rochelle M. Stiles

CRC Press
Taylor & Francis Group
Boca Raton London New York

CRC Press is an imprint of the
Taylor & Francis Group, an **informa** business

CRC Press
Taylor & Francis Group
6000 Broken Sound Parkway NW, Suite 300
Boca Raton, FL 33487-2742

© 2020 by Taylor & Francis Group, LLC
CRC Press is an imprint of Taylor & Francis Group, an Informa business

No claim to original U.S. Government works

International Standard Book Number-13: 978-0-367-35683-5 (Hardback)
International Standard Book Number-13: 978-0-367-45635-1 (Paperback)

**Visit the Taylor & Francis Web site at
http://www.taylorandfrancis.com**

**and the CRC Press Web site at
http://www.crcpress.com**

To the women and men of the Indiana Department of Natural Resources who supported this project, and to our many colleagues who gladly lent their expertise to help us understand these strange and wonderful beasts. And to the memories of Daryl Karns, Marcy Sieggreen, and Tim Halliday—colleagues and collaborators on the work described here, and cherished friends who died too young.

"One of the chief privileges of man is to speak up for the universe."

Norman Maclean (1992)

"We say of other creatures, 'Ah, they're just animals,' and they are. But we have to expand our definition of animal every time we get to know one better."

Douglas H. Chadwick (2010)

"Research is sometimes a matter of sheer serendipity, but if anything, it is more a matter of persistence, of sheer bloody mindedness."

Richard Fortey (2008)

Contents

Preamble

A single male Crawfish Frog calling in the distance at night is a lonely, plaintive sound. Locating the source of the frog is difficult; he is always more distant than his vocalizations suggest. Approach, he will go silent, and you will never find him. But on those increasingly rare nights when large numbers of Crawfish Frogs gather to form a booming chorus, males will not stop calling as you approach. Get close and the collective sound becomes so loud, rather than hear it you feel it as a reverberation juddering your brain; like standing next to Niagara Falls with your eyes closed.

Getting to know Crawfish Frogs offers a similar experience. These cryptic, obscure amphibians resist revealing their secrets. But be persistent and maybe a little unconventional in your approach and they start to show themselves. After enough of these disclosures, you begin to comprehend that these are not the animals the scientific literature said they were. You realize specialists sometimes get it wrong; that Crawfish Frogs are not the slow, clumsy animals you read about; and therefore, by inference, deserve the extinction they are steamrolling toward. Instead, you understand these animals have hit upon one of the most ingenious evolutionary inventions in the history of life on Earth—the cooption then customization of crayfish burrows to create a deeply effective method of predator defense. Further, Crawfish Frogs appear to have modified their skulls to form a shield, to strengthen this defense. Impressions of disdain turn to wonder as yet another unsuspected animal survival trick is revealed to a collective human consciousness that thought it knew it all. And you are relieved, but not comforted, to know that this discovery of life's possibility was uncovered before extinction snuffed out yet another animal novelty. A novelty, which in combination with all other evolutionary innovations, represents the realized potential of life on planet Earth.

Preface

Aside from humans, a mere handful of organisms—our fellow primates, whales, dolphins, wolves, big cats, and perhaps a few bird species such as ravens and crows—use their brains to contemplate. Only these sorts of animals have a strong enough sense of themselves and the world around them to mull, ponder, strategize, and consider risks prior to making decisions. This select group does not include the amphibians.

Here, we tell the story of a frog that took a risk and became highly specialized. Then, through circumstances beyond its control, got itself in trouble and now finds itself in deep distress. As we describe this gambling frog, we are not suggesting these amphibians were weighing the odds during the process of making this decision. Instead, we're talking about the historical process of success or failure, termed natural selection, discovered by Charles Darwin and Alfred Russell Wallace in the middle of the nineteenth century.

Darwin and Wallace's argument goes as follows. Organisms exhibit variation (if they did not, you would be unable to distinguish your father from your uncle, or a Chihuahua from a Great Dane). Because of this variation, some individuals—those better attuned to their current environment—generate relatively more offspring that will generate relatively more offspring, and so on. The wild animals and plants we experience today are the products of thousands of generations of such successes. As humans, we tend to put value judgments (winners or losers) on the outcomes of this process, but in fact success in this sense is a relative and capricious term. A species of algae that once thrived under Antarctic ice can hardly be blamed for failing in a world of global warming.

At some point in the distant past, a group of frogs became successful by seeking shelter in caves. They had proportionally more offspring that had more offspring, and so on—a term we call fitness. The fittest among this ancient lineage gave rise to today's Pickerel Frogs. A subset of their descendants then carried this tendency to occupy cryptic retreats further, and began inhabiting nooks and crannies such as tortoise burrows, stump holes, mammal runs, and abandoned crayfish burrows. These frogs ranged to feed, then returned to their shelters to rest, avoid environmental extremes, and elude predators. Being more sedentary, their bodies got heavy, and they developed defensive behaviors based more on holding their ground than fleeing. They were successful, and gave rise to today's Gopher Frogs (two species).

Some frogs in this lineage then carried this tendency toward smaller retreat sites further, and began using only crayfish burrows. These Crawfish Frogs added three twists to this evolutionary story. First, they extended this lineage's trend toward a sedentary lifestyle to its extreme by almost never leaving their burrow entrances except to breed. Because of this preference, and the defensive behaviors they developed in response, Crawfish Frogs were able to avoid predators more effectively than other ground-dwelling frogs, and because of this advantage achieved a long lifespan (which meant on average more annual reproductive opportunities for each frog). Second, as they grow, Crawfish Frogs change the shape of their skull from pointed in juveniles to round in adults. This shape-shifting means that by the time Crawfish Frogs are large enough to use burrows in predator defense, their skulls provide a shield—a sort of lid that seals off their soft body from the grasp of a predator as they face their burrow opening. Third, because crayfish dig their burrows deeper than frost penetrates the soil at mid-latitudes, these freeze-intolerant frogs survive in regions that experience winter. With this ability, Crawfish Frogs dispersed north, and expanded their range far beyond that of their nearest relatives, the Gopher Frogs. There was a cost. In exchange for long life and an expansive range, Crawfish Frogs tied their fate to the fortunes of upland crayfish and behaviors that offered advantages to burrow dwelling.

There was a second cost. The sedentary lifestyle of Crawfish Frog adults generated a more sedentary tadpole—after all, a frog that begins life as an egg, that then becomes a tadpole, that then becomes a juvenile, that then becomes an adult, is at every stage of its life the exact same animal. Slower tadpoles were less able to avoid predators and were likely disadvantaged competitors when sharing wetlands with tadpoles of other, more active species. Fortunately, Crawfish Frogs arose in an area where bison were wallowing out shallow basins to create new wetlands. It takes newly formed wetlands some time to establish a full complement of plants and animals—to approach their ecological promise*—and Crawfish Frogs may have taken advantage of this lag. By breeding in these newly formed, bison-wallowed wetlands, adult Crawfish Frogs may have given their tadpoles the advantage of reduced exposure to predators and competitors.

At this point in their history, Crawfish Frogs had it made. They owned a widespread distribution and a long lifespan; they owed their head shape to burrow defense; and, by any measure of accomplishment, were successful. But the foundation of their success rested on the fates of two vastly different types of

---

* To become a climax community in Frederic Clements' view of ecological succession (Lannoo 2018).

animals: their upland habitat was tied to the fortunes of crayfish, their wetland habitat to the destiny of bison. No other amphibian in North America was so co-dependent. It was a gutsy strategy—a true gamble. But prior to Europeans settling North America it worked. At the height of their success, Crawfish Frogs occupied much of the southeastern Great Plains and Mississippi Delta, an area encompassing what we now recognize as portions of 13 states. Every place Crawfish Frogs could inhabit, they did. Then, things changed. In 1492, Europeans returned to North America. And this time, the white man was here to stay.

Front Cover Photo: Crawfish Frog (*Rana areolata*) inhabiting its crayfish burrow. Photo by Nate Engbrecht and used here with permission.

Back Cover Photo: Crawfish Frog (*Rana areolata*) at dusk. Photo by Michael Redmer and used here with permission.

Introduction

1

On 1 April, 2010, Jen Heemeyer* (Figure 1.1) was radiotracking Crawfish Frogs exiting wetlands following their spring breeding efforts. Around mid-morning, she called MJL's cell: "Frog 160 returned to her burrow, the same one she occupied all last summer." With these two facts—that Crawfish Frogs will not only occupy a single crayfish burrow throughout the year but return to the very same burrow after leaving it to breed during the following year—Heemeyer had learned that Crawfish Frogs have a "home," a place of their own. She realized these frogs don't simply occupy any old nook or cranny, or abandoned crayfish burrow; instead, there is something more behind their habitation of these old crayfish burrows. Following up on Jen's breakthrough, we have now observed post-breeding Crawfish Frogs returning to their burrows for as many as five years in a row, a degree of site fidelity not usually seen in amphibians. Equally improbable, some of these home burrows are located over a kilometer from breeding wetlands. That's a long trek for a frog that is 10 centimeters (4 inches) long.

Crawfish Frogs (*Rana areolata*) acquired their common name from their tendency to inhabit crayfish-dug burrows. Before Heemeyer's work, the scientific literature had provided a handful of often-contradictory observations and lightly informed speculation about the nature of Crawfish Frog burrows and how they use them. Only in the past couple of decades has technology advanced enough to enable researchers to readily address such questions, and only recently has the future of Crawfish Frogs been uncertain enough to require answers.

Crawfish Frogs are "true frogs," members of the Ranidae family. This group has a widespread, global distribution, which includes temperate, tropical, and some polar regions. In North America, Crawfish Frogs are distantly related to the common and familiar Leopard Frogs,[†] Bullfrogs,[‡] Green Frogs,[§]

---

* Jen has since gotten married to Dr. Andrew Beck and taken his last name. We will use her maiden name here, because that's how we came to know her, and she used her maiden name when publishing her Crawfish Frog work.

[†] Northern and Southern Leopard Frogs (*Rana pipiens* and *R. sphenocephala*, respectively).

[‡] *Rana catesbeiana*.

[§] *Rana clamitans*.

FIGURE 1.1   Jen Heemeyer, displaying the energy and personality she brought into our lab.

and Wood Frogs.* Their closest relatives—so much so that as recently as 2001 they were thought to be the same species (Altig and Lohoefener 1983; Young et al. 2001)—are Dusky Gopher Frogs† and Carolina Gopher Frogs.‡

This is a group in trouble. Dusky Gopher Frogs were granted federal endangered species status in 2001, and if not for captive rearing programs at several dedicated zoos around the United States would now be extinct. In 2014, Carolina Gopher Frogs were petitioned for federal listing and, adding urgency to the petition, in 2014, there was evidence of a severe reduction in breeding activity. Similar in trajectory but not yet in extent, Crawfish Frogs are of conservation concern in every state where they occur. They were last seen in Iowa in 1942, and last seen in Louisiana during the 1970s until a birdwatcher photographed a single breeding male in February 2010 (Boundy, pers. comm.). In Indiana, where we work, Crawfish Frogs are listed as state endangered (Figure 1.2).

Despite following Gopher Frogs on the road to near-extinction, or at least toward costly recovery efforts, Crawfish Frogs offer optimism. They currently have a wider distribution than either Gopher Frog species and still have pockets of abundance, especially in eastern Kansas and Oklahoma. If their declines can be halted, or perhaps reversed (which, as we will see is not a difficult thing to do), the expensive and uncertain process of recovery can be avoided.

Perhaps the biggest challenge in determining the conservation status of Crawfish Frogs is they are difficult to find—they inhabit abandoned crayfish

---

* *Rana sylvatica.*
† *Rana sevosa.*
‡ *Rana capito.*

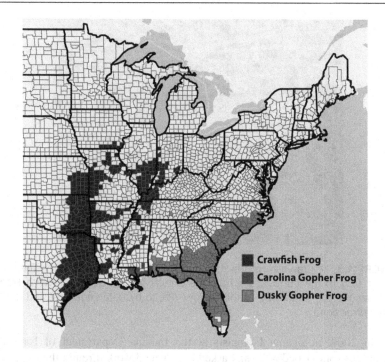

FIGURE 1.2 The historical distribution of Crawfish Frogs, Carolina Gopher Frogs, and Dusky Gopher Frogs. Note that the historical distribution of both Gopher Frog species followed the Gulf and Atlantic Coastal Plains. In contrast, the historical distribution of Crawfish Frogs tracked north from eastern Texas through Oklahoma and Kansas to form a large continuous distribution, then followed rivers and grasslands east and north across Louisiana, Mississippi, Alabama, Arkansas, and Missouri to form a second continuous distribution in the southern portion of Illinois and Indiana, and the western portions of Kentucky and Tennessee.

burrows in tallgrass prairie, and therefore are not easily detected. In 1927, the herpetologists Herman Wright and George Myers wrote, "Probably no North American *Rana* is so little known, both to its habits and its distribution" (Wright and Myers 1927). In the early 1950s, Arthur Bragg (1953) wrote, "For each of the past four years [1945–1948 inclusive] I have spent two or more months in field study of amphibians … without once observing these frogs in nature, despite being constantly on the alert for them." This begged the question, first asked over 90 years ago, whether Crawfish Frogs were truly rare and needed conservation protection and survival assistance, or instead were abundant but perceived by humans to be rare because they can be so frustratingly difficult to detect. Howard Gloyd thought he knew the answer: "This species [is] more common than is ordinarily supposed" (Gloyd 1928) (Figure 1.3).

FIGURE 1.3 A Crawfish Frog in its primary crayfish burrow. Note the tight fit of the frog to the burrow, and how the circular jawline of the Crawfish Frog matches the size and shape of the burrow. (Photo by Nate Engbrecht and used with permission.)

In 2008, non-game biologists at the Indiana Department of Natural Resources approached MJL and asked if he would work through the life history and natural history of their state endangered Crawfish Frog, and offer management recommendations in order to conserve them. MJL had just moved to Terre Haute, and was the only research-oriented academic herpetologist in Indiana living within the historical range of Crawfish Frogs. He was also the United States' Chair of the International Union for Conservation of Nature's (IUCN's) Amphibian Specialist Group, and in that capacity had just edited a large volume on the conservation status of all ~300 species of United States' amphibians (Lannoo 2005). He had the literature on Crawfish Frogs at his disposal, and had enough field experience to know which techniques would get definitive answers quickly. He agreed.

Our funding came from the U.S. Fish and Wildlife Service's State Wildlife Grant (SWG) program, which provides federal money for activities that benefit all forms of wildlife and their habitats. SWG priorities are based in part on State Wildlife Action Plans (SWAPs), which each state is required to assemble if they wish to receive federal wildlife funds. SWG projects are designed to address and solve smaller state wildlife problems before they become big, expensive, federal endangered species problems.

Our first field season began in early 2009, and as we write this in the fall of 2016 most aspects of this project have been completed. As George Schaller quipped, "If you have a long-term dataset you find out what actually happens"

(Quammen 2013). While we did not work with Crawfish Frogs as long as we would have liked, eight consecutive field seasons of intensive study is lengthier than most field projects. We received two SWG grants. The first, for three-and-a-half years plus a one-year extension, was devised to establish a basic life history and natural history database for this species. The second, for three more years, was designed to develop management techniques to maintain or augment existing populations, and to create protocols for re-establishing populations in regions where they had been extirpated.

What follows is the Crawfish Frog story, and, maybe more than a little bit, our story: How a small mom-and-pop lab stocked with gifted graduate students and networked to exceptional professionals willing to assist came to understand the critical, linchpin aspects of the lives of these frogs—the factors that determine whether they live or die. We understand that knowing these facts will not inevitably pull Crawfish Frogs back from the brink, but they do provide American citizens with the knowledge to do so, should they decide this species has a right to exist.

As you read through the following stories, you may wonder about the names or designations we assign to frogs. There was little rhyme or reason to our nomenclature except for ease of communication. Most frogs were anonymous, known to us only by their cohort toe clip markings (e.g., Nate's Pond, 2009) or their pit tag numbers (e.g., 4b08253605), which we never memorized but could look up. The subset of frogs we specifically refer to below are, in all but one case, animals that had radio-transmitters in them. Jen had two designations for these frogs. One was simply a sequential number based on their surgical order. For example, Frog 26 was the 26th frog Jen operated on to implant a transmitter. Most often, we began referring to these frogs by the frequency of their implanted radio-transmitter. For example, we started calling Frog 6, 139 (based on its radiofrequency of 150.139), Frog 3 became 060, Frog 5 became 080, Frog 8 was 160, and so on. When we replaced the radio-transmitters, we usually continued to call the frog by its original frequency number. When 139's original radio died and was swapped out for a radio with the frequency 150.660, we continued to call him 139. There were two reasons for this. First, we had come to know the frog by its original number, and when we mentioned the number everybody knew which frog we were referencing. Second, we recycled transmitters with dead batteries, sending them back to Holohil to be refurbished. Holohil replaced the batteries but they did not change the frequency of the transmitter. Because each of our transmitters had a unique frequency, when we swapped out transmitters in the frogs the replacement transmitter had to have a different frequency than the original transmitter. If we had always referred to each frog by its current frequency, Frog 139 in 2009 would have become Frog 660 in 2010, and we would have had to distinguish Frog 660 in 2009 from Frog 660 in 2010, which were different frogs. This would have become a nightmare. By keeping the original designation, Frog 139 was always Frog 139 to us, no matter the frequency of his current transmitter.

A small subset of Jen's numbered frogs then acquired a more descriptive name, for example, instead of being identified by his radio frequency, 279, Frog 26 became Romeo; Frog 33 became Juliet; we called Frog 53, the big female from Nate's Pond who had survived the big 2009 prescribed prairie burn, Corner Burn Chick; Frog 65, living in a neighbor's pasture, became Private; and Frog 47, near the hunting sign-in kiosk, somehow became Sugar Shack. Romeo and Juliet acquired their names from their habit of burrow sharing (twice, for two days each in 2010). Corner Burn Chick, Private, and Sugar Shack were named for the locations of the first primary burrows we found them inhabiting.

Who Are You?

2

Crawfish Frogs are "true frogs," members of the anuran family Ranidae. Ranids are a cosmopolitan family, with a worldwide distribution excepting Antarctica and Greenland (where no amphibians of any kind occur—yet). With over 1,000 described species, ranids rank among the most successful of frog families. Nearly one of every seven living frog species is a ranid, and globally there are about twice as many ranid frog species as there are species of salamanders.

DNA evidence supports the fossil record, and indicates that ranids arose ~135 million years ago and spread across the supercontinent Gondwanaland. Ranids continued to diversify following the breakup of Gondwanaland, ~102 million years ago. By ~57 million years ago, the subfamily Raninae arose. Within this subfamily, ~31 million years ago, New World ranids appeared, invading—like humans eventually would—from eastern Asia over the Bering Land Bridge (Macey et al. 2006). The main lineage was termed the *Novirana* group, and it continued to diversify (Bossuyt et al. 2006). First, the ancestors of Bullfrogs and their relatives (*Aquarana*) split off. Then, the ancestors of Leopard Frogs and their relatives (*Sterirana*) split off. Finally, a group that now includes Plains Leopard Frogs and Southern Leopard Frogs (*Scurrilirana*) split off. The remaining group (*Nenirana*) includes the species that gave rise to Pickerel Frogs, Gopher Frogs, and Crawfish Frogs (Hillis and Wilcox 2005; Pyron and Wiens 2011) (Figure 2.1). With the exception of Pickerel Frogs, which have dispersed as far north as eastern Canada, this group arose and today occurs entirely within the boundaries of the continental United States.

Some details. Pickerel Frogs have been called the most cave-adapted frog in North America (Parris and Redmer 2005), and represent the ancestral *Nenirana* form. Coleman Goin and Graham Netting* suggested that this lineage arose in east Texas. As these frogs dispersed, one branch spread east along the Gulf Coast, crossed the region that was to become the Mississippi Delta, continued along the Gulf Coast into peninsular Florida, and then kept dispersing northeast along the Atlantic Coast to southern North Carolina.

---

* Goin and Netting (1940); Smith and Minton (1957) agreed.

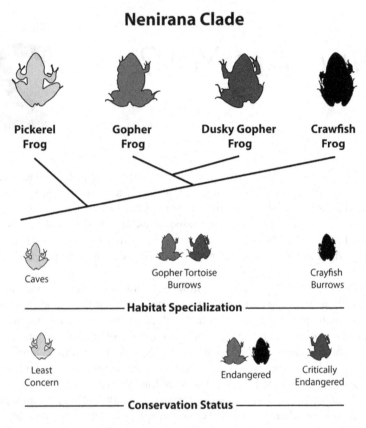

FIGURE 2.1 The evolutionary relationships among the four species within the subgenus *Nenirana*. Pickerel Frogs are slender-bodied, the most basal species, often inhabit caves, and are of least conservation concern. The lineage then splits into the heavy-bodied, derived Gopher Frogs and Crawfish Frogs. Gopher Frogs are named for their tendency to inhabit Gopher Tortoise burrows, while Crawfish Frogs are crayfish burrow dwellers. All three species of Crawfish and Gopher Frogs are of conservation concern, and Dusky Gopher Frogs would be extinct if not for the husbandry efforts and talents of zoo and aquarium biologists and veterinarians.

Today, we recognize this lineage as the Gopher Frog group. The second branch of the east Texas lineage, aided by the resistance to freezing afforded by crayfish burrow dwelling, spread north along the eastern Great Plains across eastern Oklahoma into southeastern Kansas. These frogs also went east into northwestern Louisiana, across Arkansas, and through Missouri (avoiding the Ozark Highlands) into southeastern Iowa. They managed to cross major rivers such as the Arkansas and Missouri, then crossed the Mississippi to enter the southern half of Illinois, southwestern Indiana, western Kentucky and Tennessee, and parts of central Mississippi into Alabama, where they nearly rejoined the Gopher Frog lineage (indeed, Wilfred Neill suggests they did (Neill 1957)). These frogs became our Crawfish Frogs, with two subspecies separated on a northwest to southeast line running through east-central Oklahoma and across Arkansas to the Mississippi River. All frogs in Texas, southeastern Oklahoma, southwestern Arkansas, and Louisiana are considered the more ancestral Southern Crawfish Frogs (*Rana areolata areolata*). All frogs to the north of this line west of the Mississippi River, as well as all frogs east of the Mississippi River, are recognized as the derived Northern Crawfish Frogs (*Rana areolata circulosa*). Examining the map of Crawfish Frog distribution (Figure 1.2) (Lannoo 2005) and taking into account climate and vegetation, it appears that Crawfish Frogs are limited to the south by the Gulf of Mexico, to the west by the arid Plains, to the north by extreme winter cold, and to the east by the dense summer shade of the Eastern Deciduous Forest. Within this distribution, Crawfish Frogs are found only in areas where upland crayfish are able to dig burrows, so they are not found in the rocky Missouri Ozarks or in the Kankakee Sands region of northern Indiana. They also do not reliably occur in low-lying areas that regularly flood.

Despite their widespread distribution across territory long settled by European pioneers, the United States was three-quarters of a century old before scientists discovered Crawfish Frogs. In areas west of the gateway city of St. Louis, Native Americans, Mountain Men, Voyageurs, and nineteenth-century homesteaders must have heard the snore-like spring breeding calls of Crawfish Frogs, which will carry up to two kilometers under favorable acoustic conditions. In 1928, Howard Gloyd noted that the Crawfish Frogs' "cry, if not the frogs [themselves, were] familiar to almost everyone who listens to frog voices in the spring" (Gloyd 1928). People must have sat by their campfires, heard these calls, and wondered "What the ...?" (Wright recounts Florida pioneers hearing the similar call of Gopher Frogs and calling it "the monster" (Wright 1932)). Such a strange organic sound—distant, mysterious, and ominous—coming after sunset and continuing long into the night, from animals you never saw in daylight, indeed must have unsettled the settlers.

There may continue to be adjustments to how scientists classify Crawfish Frogs (for example, the results from DNA studies may support species-level designations for the two currently recognized subspecies), but the biggest question now is biogeographical—what will future data tell us about Goin and Netting's notions of the rise and spread of the Crawfish Frog/Gopher Frog group? If their scenario holds, it is impressive that Crawfish Frogs, which are not as aquatic as Leopard Frogs and will drown, managed to cross several deep rivers, including the formidable Mississippi, to achieve their expansive pre-European-settlement distribution.

Another curious aspect of the Crawfish Frog story is how and where they finally revealed themselves to science. In the first half of the nineteenth century, no frontier city was larger, more cosmopolitan, or had as many people moving through it than St. Louis. Crawfish Frogs must have occurred in many locations in and around this frontier metropolis and cultural hub, but somehow the scientists on numerous expeditions* missed collecting them. Instead, Crawfish Frogs were discovered in Texas by the federal Mexican Border Survey team, half a century after Lewis and Clark used St. Louis as their gateway to discovery. The story goes like this.

In 1850, the federal government, through its Corps of Topographical Engineers, began five major surveys to establish a series of transcontinental railroads, from the Great Northern Line, which would eventually run from the Twin Cities to Portland in the north, to the Southern Pacific Line, which would run from New Orleans to San Diego in the south. Field teams of botanists, zoologists, and ethnographers accompanied Corps' geographers. They collected specimens and sent them to the newly established Smithsonian Institution for identification, with type specimens of new species deposited in its National Museum. Similarly, the federal government sent survey teams of Topographical Engineers to explore the boundaries between the United States and Canada, and the United States and Mexico. In the Mexican survey—formally, the U.S. Mexican Boundary Survey—J. H. Clark collected the first Crawfish Frog known to science. The specimen was cataloged as United States National Museum (USNM) No. 3304.

At that time, Spencer Baird was the Assistant Secretary of the Smithsonian and, as curator of collections and operations at the National Museum, in charge of all animal specimens being shipped back East from these federal surveys. In 1852, Baird and his herpetologist, Charles Frédéric Girard, published the first description of "*Rana areolata*" based on Clark's specimen, in an article entitled "Characteristics of some new Reptiles in the Museum of the Smithsonian Institution." Baird and Girard (1852, p. 173) wrote as follows:

RANA AREOLATA, B. and G. Head very large, sub-elliptical; snout prominent, nostrils situated half way between its tips and the anterior rim of the eyes, which are proportionately large. The tympanum is spherical, and of medium size; its central portion is yellowish-white, whilst its periphery is black. The body is rather short and stout; the limbs well developed; the fingers and toes very long without being slender. The ground color of the body and head is

---

* For example those of Meriwether Lewis and George Rogers Clark, Stephen Long, and Zebulon Pike.

yellowish-green, marked with dark brown. Besides there are from thirty to fifty brown areolæ, margined with a yellowish line. The upper part of the limbs is of the same color as the body, but instead of areolæ, transverse bands of brown are seen on the hind ones. The lower part of the head and body is yellowish, with small dusky spots along the margin of the lower jaw, and under the neck.

A specimen three inches and a half long was found at Indianola [Texas], and a small one on the Rio San Pedro of the Gila.*

Three years after Baird and Girard published their description, John Le Conte attempted to clarify the taxonomic standing of the accumulating collections and the confusing nomenclature beginning to arise among North American ranid frogs. Given his sentence structure, one could easily be forgiven if they doubted the ability of Le Conte to clarify anything. But his grammatical short-comings did not cripple his career; in 1876, he became the President of the University of California, Berkeley, a position he held for five years. In 1855, Le Conte wrote:

On account of the numerous errors which have been introduced into that part of the Herpetology of our country, which treats of the frogs and allied animals, I have been induced in this short memoir, to reduce such as I have been able to obtain, to some better order. I offer but a descriptive catalogue. To this are joined all the synonyms which I have been able to collect from works which could be consulted in this city.

In his report, Le Conte provides the first description of *"Rana capito,"* collected from Georgia.

R. CAPITO.

Above very rough, dark grey or slate-color speckled with black with six rows of roundish spots on the back; sides speckled and irregularly marked with spots of the same form and color; from the orbits to beyond the middle of the body runs a broad raised line or cutaneous fold; and another from the corner of the mouth to the insertion of the arm. Beneath smooth, yellowish white, speckled, spotted and varied with dusky; top of the head coarsely punctured, back and sides tuberculous. Head very large, broad and blunt, a deep concavity between the nostrils and the eyes. Irids golden mixed with black. Tympanum the color of the body. Lower jaw with a small protuberance or point resembling a tooth. Arms and legs above grey, speckled and barred with black; beneath yellowish spotted and varied with dusky, the yellowish

---

* According to Altig and Lohoefener (1983) the record from Rio San Pedro, a tributary of the Gila River, in Arizona, is erroneous.

color more decided at the axillae and groins. Hind part of the thighs granulate. Fingers slightly palmate at the base, the first longer than the second. The second toe twice as long as the first.
Length 4·2 in., width of the head at the corners of the mouth 1·5 in., arm 1·87, leg 4·75, thigh 1·1 in., tibia 1·45, foot 2·2.
Inhabits Georgia in the ditches of the rice fields.*

Twenty years later, in 1875, Edward Drinker Cope, one of Baird's students at the Smithsonian and a member of the irreverent "Megalotherium Club" (Lannoo 2018), published his *Check-List of North American Batrachia and Reptilia* (Cope 1875). In it, he listed the specimens described by Baird and Girard (USNM 3304) and Le Conte (USNM 5903) as subspecies, as follows:

> *Rana areolata*, Baird and Girard, subspecies *areolata*, Bd. Gir., U.S.
> Mex. Bound. Surv., 28, p. xxxvi, Figs. 11–12. Texan district.
> *Rana areolata*, Baird and Girard, subspecies *capito*, Le Conte, Proc.
> Acad. Phila., 1855, p. 425. Floridan district.

Cope's subspecies designations appear to have been based on one actual and one perceived difference in coloration. He noted that *areolata* have light rings surrounding the dorsal "areolæ," (true), and *capito* have "speckled, spotted and varied with dusky" markings on their belly (not a definitive feature, we now know that many *areolata* also have speckled bellies).

In 1878, three years after Cope published his *Check-List*, Frank Leon Rice and Nathan Smith Davis described a new form of *Rana areolata* collected by Dr. Elias Francis Shipman (Resetar and Resetar 2015) from Benton County, Indiana, and deposited in the Chicago Academy of Sciences (CAS 160) (Jordan 1878). Their description reads:

> *R. circulosa*, Rice and Davis (sp. nov.). Hoosier Frog. Head broad; body, head and sides with the ground color largely predominating, and with narrow rings of a greenish slate color, which become larger and more irregular posteriorly; hind legs black, crossed with irregular lines of yellowish slate color; fore limbs similarly marmorate; tympanum black with pale ring; below chiefly yellowish white; toes very long; size medium; L. 3 ½. Benton Co., Indiana, lately discovered by Mr. E. F. Shipman (Figure 2.2).

---

* Altig and Lohoefener (1983) wrote: "Harper (1935) assumed Riceboro, Liberty County, Georgia was the type-locality since Le Conte had a plantation there; Cochran (1961) listed this locality. Neill (1957) pointed out that collectors sent Le Conte specimens and Liberty County was not necessarily the type locality. He agreed with Schwartz and Harrison's (1956) doubts that the holotype (USNM 5903) was actually the specimen described by Le Conte (1855), but Harper (1935) attributed size differences to shrinkage."

FIGURE 2.2 Nate Engbrecht holding the original Hoosier Frog, Chicago Academy of Science specimen 160, collected by Francis Shipman and described by Rice and Davis in 1878.

Over the next 90 years there were adjustments to this taxonomy, so that by 1970, Doris M. Cochran, the Curator of Amphibians and Reptiles at the National Museum, and Coleman J. Goin recognized two subspecies of Crawfish Frogs: Baird and Girard's *Rana areolata areolata* and Rice and Davis' *Rana areolata circulosa*. They also recognized three subspecies of Gopher Frogs: Le Conte's *Rana capito capito*; *Rana capito aesopus*, collected by T. H. Bean from Micanopy, Alachua County, Florida (USNM 4743); and *Rana capito sevosa*, based on a specimen collected by Percy Viosca Jr. near Slidell, St. Tammany Parish, Louisiana (Carnegie Museum #16809; initially described by Goin and Netting in 1940*).

In 1983, Ronn Altig and Ren Lohoefener sank *capito* into *areolata* and considered all five of Cochran and Goin's subspecies to be couched within *areolata* (Altig and Lohoefener 1983). Altig and Lohoefener's assessments appear to have been based primarily on Wilfred Neill's analysis of body coloration (Neill 1957), which we now know varies between sexes, seasonally within populations, and across populations. In 2001, Jeanne Young, Brian Crother, and J.D. McEachran used genetic data to once again (and presumably finally) separate *areolata* from *capito*. They also collected evidence to erect *sevosa* to full species status, which created three Crawfish Frog/

---

* The first specimen was collected by Percy Viosca Jr. on 11 April, 1926 (Carnegie Museum #16809). See also Wright and Wright (1942).

Gopher Frog species—*areolata*, *capito*, and *sevosa*—which is the taxonomy we recognize today.*

The debate surrounding Crawfish Frog/Gopher Frog scientific names was repeated for their common names, which scientists also standardize. In the process of describing Crawfish Frogs, in 1933, Albert and Anna Wright listed synonymous common names as follows: "Northern Gopher Frog, Gopher Frog, Texas Frog, Florida Frog, Crayfish Frog, and Hoosier Frog." Enter Francis Harper.[†] Of all of the early twentieth-century biologists working on Crawfish Frogs, we would have most liked to meet Francis Harper. He was never affiliated with a prestigious eastern institution, yet he had the confidence to challenge the Smithsonian's Leonard Stejneger and the legendary George Albert Boulenger (1920). While arguing for what is now the modern taxonomy of Crawfish and Gopher Frogs, Harper tackled the issue of common names and in the process coupled the common name "Crawfish Frog" to the scientific taxa "*areolata*" (Harper 1935). Harper was not subtle, and presumably had no qualms about throwing Thompson, Gloyd, and the Wrights (and later Hobart Smith and Phillip Smith—both of whom should have known better) under the bus:

> The evidence at hand indicates that the range of *Rana capito* lies wholly within [not quite correct (see Goin and Netting 1940, p. 154)], and the range of *R. areolata* wholly without, that of [the Gopher Tortoise] *Gopherus polyphemus* … Just as *capito* owes its common name to the Gopher Turtle with which it associates, so "Crawfish Frog" is a fitting name for *areolata*, by reason of its appropriation of the crustacean's burrows for its own habitations. "Northern Gopher Frog," employed by some authors for *areolata* is scarcely a suitable name for a species that is not known to have any contact with the Gopher Turtle.

Despite the factual power of Harper's arguments, it took another 18 years (1953) before "Crawfish Frog" was adopted by Karl Schmidt, representing the Society for the Study of Amphibians and Reptiles, as the official common name for *Rana areolata*.[‡] Paul Samuelson got it right when he said, "Funeral by funeral, theory advances."[§]

---

* Along with our colleagues D. Saenz and T. Hibbitts, we have recently published a paper demonstrating the difference in call characteristics between Crawfish Frogs and Gopher Frogs, supporting the decision to separate these two taxa into separate species (see Lannoo et al. 2018).
† Harper (1935, p. 81) writes: "Wright and Wright (1933, p. 150) show a clear understanding of the distributional facts by excluding *areolata* from the southeastern states, but still refrain from substituting *capito* for *areolata*."
‡ We thank Brian Crother for guidance here, in a response to MJL's query on 8 April, 2015.
§ http://www.brainyquote.com/quotes/quotes/p/paulsamuel205550.html.

Today, new taxonomic debates involve the Crawfish Frog/Gopher Frog group, but are not centered on them. For example, Daryl Frost and all three herpetological societies in North America (the American Society of Ichthyologists and Herpetologists, Society for the Study of Amphibians and Reptiles, and the Herpetologist's League) consider Crawfish Frogs, Gopher Frogs, and most other species of North American ranids to be within the genus *Lithobates* (where the species name becomes *areolatus* to provide agreement with the Latinized genus name), while Dave Hillis and Tom Wilcox, and the team at AmphibiaWeb (www.AmphibiaWeb.org), consider them to be valid *Rana* (with the species name remaining *areolata*). After recently reviewing this controversy, we consider the arguments of Hillis and his colleagues (Pauley et al. 2009) to be more compelling, and have decided to follow them here.

It would be best if all scientists agreed on what to call these animals, but a designated name hardly leads to a deeper understanding of a species' behavioral tendencies and ecological relationships; nor does it provide additional insights into effective conservation strategies. As MJL's first mentor, University of Chicago-trained zoologist Dick Bovbjerg, used to say: "So you know its name. What else do you know?"* We trust the insights we offer here will generally hold, no matter which names future scientists use to designate this species.

---

* Names, in fact tell you a lot. As you descend the taxonomic classification scheme—Kingdom, Phylum, Class, Order, Family, Genus, Species—you are summarizing a species' evolutionary history, with each rank descending from its preceding rank. For Crawfish Frogs, this summary goes like this: Kingdom Animalia, Phylum Chordata, Class Amphibia, Order Anura, Family Ranidae, Genus *Rana*, Species *areolata*.

Mr. Peabody's Coal Train

3

Then the coal company came with the world's
largest shovel
And they tortured the timber and stripped all
the land.
Well, they dug for their coal 'til the land was
forsaken
And they wrote it all down as the progress
of man.
And Daddy won't you take me back to
Muhlenberg County
Down by the Green River where Paradise lay.
Well I'm sorry my son, but you're too late in
asking
Mr. Peabody's coal train has hauled it away.
(John Prine 1971) Atlantic Records

John Prine's lament epitomizes the environmentalist attitude toward visibly destructive resource extraction, and we will not downplay the impacts such activities have on natural ecosystems and their landscapes, including the area Prine describes around the lost town of Paradise, Kentucky (Stiles et al. 2016a). But, as with all natural processes, even destructive resource extraction is nuanced. The work we did on Crawfish Frogs opened our eyes to this subtlety. Specifically, we came to realize that the ultimate effect of previous surface coal mining at our study site was to create a landscape that now resembles the historic (pre-European settlement) landscape, bringing human impacts in this region full circle. Amphibians and reptiles have responded in kind to this newly formed historical ecosystem assemblage, and as we will see this example offers a model for future reclamation and restoration activities.

Working out of the University of Chicago, in the early 1930s, Edgar Transeau (1935) mapped the historic prairie habitats of North America, with an emphasis on the eastern extension known as the prairie peninsula. His map, inelegant by today's standards (the black of the prairie masking important political boundaries and landscape features), nevertheless highlights the presence of isolated prairies along the edges of the peninsula in southern Missouri, southern Illinois, southwestern and north-central Indiana, southwestern Michigan, and southern Wisconsin (Figure 3.1). The historic fauna of these "pocket" prairies presumably resembled the fauna of the larger grassland peninsula, with some reduction of species based on the size of the isolate (MacArthur and Wilson 1967). In southwestern Indiana, the memory of these settlers' initial surroundings is retained in the current names of towns (Prairietown, Prairie City, Hoosier Prairie, Walnut Prairie) and natural areas (Prairie Creek).

While habitat destruction, disease, invasive species, pollution, and the collection of wild-caught animals to supply the pet trade are all factors known to impact amphibian populations (Collins and Storfer 2003; Stuart et al. 2004; Daszak et al. 2005), habitat loss is likely the most important global cause of amphibian declines (Collins and Storfer 2003; Bradford 2005; Gallant et al. 2007). With this in mind, we recall that Midwestern settlers first plowed the grasslands of the prairie peninsula for agriculture in the mid- to late nineteenth century, and that this landscape remained in agricultural production for over a century. Not long after the settlers arrived, geologists recognized

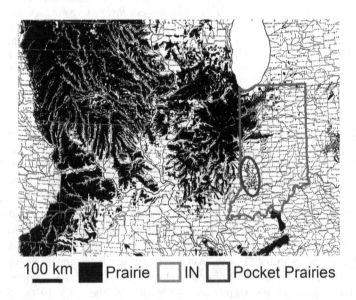

100 km ■ Prairie ☐ IN ☐ Pocket Prairies

FIGURE 3.1 Transeau's (1935) classic map showing the prairie peninsula extending eastward through the central Midwest. We have added highlights to show both the political boundaries of Indiana and the cluster of historic isolated pocket prairies that characterized the region encompassing our study site. Used with permission of the Ecological Society of America and John Wiley and Sons.

the presence of coal—deposits that are now called the Illinois Basin—in the region that encompasses a vast area of the central Midwest running from the Quad Cities along the Iowa/Illinois border east to Joliet, southeast to western Kentucky (including Muhlenberg County), west to St. Louis (home of Peabody Coal headquarters), and back north to the Quad Cities. The Illinois Basin is bowl-shaped, and coal is efficiently removed from its shallow margins using surface mining techniques (Figure 3.2).

The history of coal extraction in southwestern Indiana is over a century old. Prior to 1977, surface mining was a crude business, with overburden piled on the surface—one shovel-full dumped adjacent to the previous and so on in a pattern resembling an upside-down egg carton—and planted over by state reclamation biologists with acid-tolerant white pine seedlings. The pit was then kept open and allowed to fill with groundwater to become a small, steep-sided recreational lake. The net effect of this activity, conducted without much regulatory oversight through much of the twentieth century, has punctuated the common Midwestern corn belt landscape with an occasional lake-and-pines look reminiscent of Minnesota's northwoods or a Rocky Mountain river valley.

Reclamation procedures were legislatively revised in 1977 with the implementation of the federal Surface Mining Control and Reclamation Act (SMCRA). Under SMCRA guidelines, habitats destroyed as a result of coal mining had to be restored to their previous use, a historical use, or to

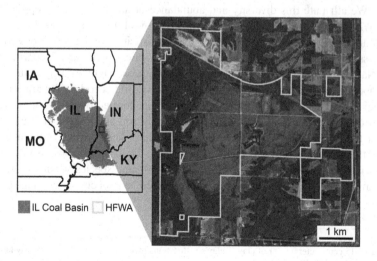

IA

IL   IN

MO

KY

IL Coal Basin   HFWA

1 km

FIGURE 3.2 The location of our study site, Hillenbrand Fish and Wildlife Area-West, with reference to the Illinois Coal Basin (left) and the detailed boundaries of this Indiana natural area (right). Moss Lake is in the center of the photo; most of our work was conducted in the western portion of this property.

a standard that is equally or more economically productive. In southwestern Indiana, post-SMCRA reclaimed mine sites are now being used as agricultural fields, housing developments, and recreational sites. When recreational sites were designated for wildlife habitat they were often planted to grasslands, not because the restoration biologists had historic pocket prairie habitats in mind, but because tree roots could not penetrate the compacted, hard pan soil produced by the re-contouring of the overburden fill by heavy earth-moving equipment.*

As a component of our Crawfish Frog project, we documented the diversity and abundance of all amphibians and reptiles at our study site, Hillenbrand Fish and Wildlife Area-West (HFWA-W), in Greene County, Indiana (Lannoo et al. 2009; Kinney et al. 2010; Terrell et al. 2014b; Stiles et al. 2016a). From 1976–1982, HFWA-W was a 30-meter-deep, open-pit strip mine, where we assume (although cannot prove) that no amphibian or reptile species occurred. In the first seven years of our study, we documented 14 species of amphibians and 21 species of reptiles (Appendix I). Two of these species—Crawfish Frogs and Kirtland's Snakes[†]—are listed as Endangered in Indiana; three species—Eastern Box Turtles,[‡] Blanchard's Cricket Frogs,[§] and Rough Greensnakes[¶]—are listed as Species of Special Concern. Several species at HFWA-W are impossibly abundant, and all amphibian and reptile species appear to be successfully reproducing and recruiting new breeding adults into their populations.

We attribute this diversity and abundance of amphibians and reptiles at HFWA-W to at least four factors: 1) the creation of a variety of wetland and upland habitat types; 2) the presence of nearby offsite source populations with access to the site; 3) ecosystem management practices by Indiana Department of Natural Resources land managers; and 4) railroad rights-of-way managed as open areas that we suspect are serving as wildlife corridors for grassland species. At HFWA-W, most wetland basins were formed by depressions created during the post-mining reclamation, or by soil slumping that occurred after the restoration. While these basins may have been formed unintentionally, their role in creating amphibian and reptile diversity provides insight into how restoration biologists can favor these non-game species when restoring reclamations. The result, as we have documented, is a species assemblage that

---

* This was especially true in the late 1970s and 1980s, when overburden was capped with a thin (~15 centimeters) layer of clay. Today's restorations incorporate a much thicker clay layer (~2 meters) and will support trees.

[†] *Clonophus kirtlandii.*

[‡] *Terrepene carolina.*

[§] *Acris blanchardi.*

[¶] *Opheodrys aestivus.*

approximates the pre-settlement herpetological community (Lannoo et al. 2009; Kinney et al. 2010; Terrell et al. 2014b; Stiles et al. 2016a).

From this perspective, we can see that the subject of John Prine's lament—the loss of Paradise (both literally and, we suppose, figuratively)—is not a necessary and universal byproduct of surface coal mining. The land currently being represented by HFWA-W was in a region that historically supported pocket prairies. These prairies were destroyed 150 years ago, after European settlers introduced Western agricultural practices to the region, then agricultural practices were necessarily eliminated by the introduction of surface coal mining. The process of reclaiming these mine sites produced a degree of soil compaction that favored herbaceous over woody vegetation, creating areas of grassland mimicking the historic pocket prairies. Further, the railroad rights-of-way, which are managed as open areas (trees are trimmed and low-growing vegetation promoted), provided wildlife corridors that enabled ground-dwelling animals such as amphibians to move back into these sites. Today, HFWA-W supports a herpetofaunal diversity that rivals any comparably sized property in Indiana. The bottom line: our data suggest that areas once stripped of their ecology but since reclaimed, restored, and properly managed can be colonized by amphibians and reptiles to produce successfully breeding and presumably sustainable populations, including one of the largest populations of Crawfish Frogs east of the Mississippi River.

Shelter from the Storm

4

For more than 60 years—from their initial description by Baird and Girard in 1852 to Mary Dickerson's *The Frog Book*, published in 1906 (Dickerson 1906)—the only two things known about Crawfish Frogs were what they looked like and where they had been collected (at that time, Texas, Illinois, Indiana, and Missouri [with the Georgia record of *capito* erroneously included and the Florida record of *aesopus* correctly excluded]). In 1911, Julius Hurter, working in Missouri, was the first to add ecological information—data that began to clarify the habitat and behavioral characteristics of Crawfish Frogs (Hurter 1911). The following facts about Crawfish Frog burrow use, as stated by Hurter, are essentially correct:

> The [Crawfish Frog-occupied] holes seemed to be abandoned crayfish holes, very likely widened by the present owners.
>
> The inhabited holes are easily recognized as the entrance as well as a little platform in front of it worn smooth. Here the frog watches for its prey. As soon as it hears an unusual noise or sees someone it creeps back into the hole.
>
> All the frogs that we secured that day were not deeper down than we could reach with our hands—about 18 inches. Curiously enough they made no attempt to recede farther when we reached for them. Had they dropped to the bottom we never could have captured them, as some of the holes were three or four feet deep with water at the bottom.

Four years after Hurter's paper was published, Crystal Thompson and her sister, the University of Michigan herpetologist Helen Thompson Gaige, traveled to Olney, Illinois (100 kilometers southwest of our Indiana study site) to investigate Crawfish Frogs (Thompson 1915). The facts and conclusions drawn by these pioneering women field biologists added immensely to our knowledge of Crawfish Frog habitat and behavior, as follows.

> The region is a particularly favorable habitat for *Rana areolata* because of the presence in great numbers of the crayfish burrows, in which, as reported by Hurter, this species makes its home. These burrows are quite generally distributed in the meadows and cultivated fields but not in the sandy areas.

Their distribution apparently bears no relation to the ponds and streams, which is to be expected since the ground water level is above the hard pan and it is only necessary for the crayfish to extend their burrows into the sub-soil to insure a constant supply of water.

The old burrows occupied by crayfish were entirely without chimneys, and were approximately round at the entrance, which had a diameter of about three inches. The entrance was more or less overhung with grass and at one side was a small bare space about six inches in diameter.

Hurter says of the burrows inhabited by frogs, 'The inhabited holes are easily recognized as the entrance as well as a little platform in front of it is worn smooth.' Our observations did not bear out this statement. We were quite unable to distinguish by external appearance the old holes occupied by crayfish from those inhabited by frogs. The openings were of approximately equal size and overhung with grass, and the platforms were nearly always present ... Only rarely was an old hole found without a platform and in such cases when dug out it was found to be either entirely abandoned or to contain a crayfish. On the other hand many holes with well-developed platforms contained crayfish. After excavating for some distance we were able to determine holes occupied by frogs by the slightly smoother appearance of the walls and by the presence of beetle fragments adhering to them. [This paragraph corrects Hurter on a minor point. Certain crayfish-inhabited burrows will also have a bare spot in front, where nocturnal crayfish place their chelipeds (claws) while waiting to ambush passing prey. Indeed, the two of us cannot distinguish between crayfish-occupied burrows and Crawfish Frog-occupied burrows; our colleague Nate Engbrecht can, although he has never been able to explain how.]

Apparently when alarmed the frogs do not ordinarily descend far into the burrows, for they are plowed out in numbers and the ground in that region is only plowed to a depth of about three inches. When one attempts to dig them out, they descend farther into the holes, the exceptionally powerful hind limbs and the extent to which they can distend the body serving to secure them so firmly that they could be mutilated before being dislodged [indeed, we have observed such a mutilation, detailed below].

The frogs so nearly approximate the size of the holes that the rubbing of their soft bodies probably tends to smooth the walls, and the longer a burrow has been occupied by a frog the more shallow it becomes by reason of accumulated debris and the less liable to contain water. During the spring months, however, the water stands at the very surface of the ground and all the holes are practically filled at that time.

[Crawfish Frogs are] very agile in seeking the concealment of their burrows when alarmed.

In 1928, Howard Gloyd (1928) writing in the singular, added two critical pieces of information (italicized by us) about the use of crayfish burrows by Crawfish Frogs, although the importance of his observations went unnoted for over 80 years:

It was much more wary and difficult to catch than any other species of the vicinity, and perhaps this habitual shyness, and the fact that *it remains quiet throughout the day and in or near its burrow all the year except during the breeding season*, may account for its being so little known.

Mr. Clanton *found a specimen in a crayfish hole ... August 12. It was in a somewhat damp depression in a pasture more than a quarter of a mile from the nearest spawning pond.*

Several authors since Hurter, Thompson, and Gloyd have suggested variations or contradictions to the Crawfish Frog life history and natural history that these early naturalists detailed. Not all of them have been correct.

From a conservation perspective, a critical question for us became whether Crawfish Frogs are obligate crayfish burrow dwellers, or whether they also use other burrow types, such as mammal runs. Our first order of business was to establish the nature of their burrow use—in short, we asked what kind of burrows do Crawfish Frogs use, how do they use them, and why? (Figure 4.1).

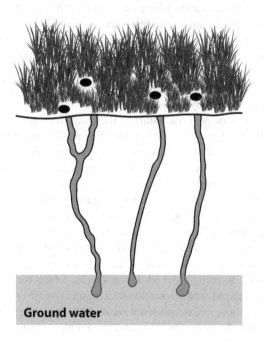

Ground water

FIGURE 4.1   A schematic drawing of one double-entrance and two single-entrance crayfish burrows with their chimneys eroded away, in grassland. Note that crayfish dig these burrows down to the water table, and then excavate a chamber.

4.1 HOME SWEET HOME

We decided the best way to determine burrow use by Crawfish Frogs would be to track frogs to burrows, and then follow their behaviors while inhabiting them. Absent snow, mud, or any other substrate likely to make an imprint, tracking small vertebrates such as frogs involves radiotracking. Basically, radiotracking involves attaching onto or implanting into an animal a (radio) transmitter that emits a signal that is then detected by an appropriately tuned receiver. We purchased our transmitters from Holohil Systems Ltd. (Ontario, Canada). Our transmitters were designed so that each was set to a different frequency, and pulsed in proportion to their temperature (which meant we could measure each frog's temperature every time we located it).

We tried both attaching the radio-transmitter to a belt and securing the belt around the frog's waist, and surgically implanting the transmitter into the frog's abdominal cavity. While it might seem that surgery is the most disruptive approach, because Crawfish Frogs move through thick, tallgrass prairie where anything external and artificial, such as a radio-transmitter, can get caught and hang them up, after some trial and error we determined that surgical implants were less disruptive to the frog, and therefore offered us the best opportunity to document their natural behaviors. Surgery was no problem; from previous research, MJL had performed hundreds of surgeries, and Jen Heemeyer was a quick study.

We collected Crawfish Frog adults as they were leaving wetlands following breeding. We brought each frog into our lab, anesthetized them, implanted a radio-transmitter, woke them, allowed them to recover, then took them back out to HFWA-W, where we released them under vegetative cover outside the drift-fence where we'd captured them.

Some details. In the lab, we anesthetized each frog by placing it in half a liter (roughly a pint) of water with its nostrils out of the water. We added 0.1 gram (a concentration of 50 milligrams/liter) of the anesthetic MS-222 (tricane methanesulfonate*); since MS-222 is acidic, we tracked the pH of the water bath. Once the anesthetic was added, we waited until the frog was completely out—unresponsive to being touched. Occasionally, a frog would not become completely anesthetized, and we would add a second dosage of MS-222. After the frog was unresponsive, we placed it on its back on a sterilized dissecting tray, and covered its body except for its abdomen with wet paper towels, which we kept wet during the surgery. Using a scalpel, we then made a small <3-centimeters (1-and-a-half-inches), midline, vertical incision through the skin,

---

* Sigma-Aldrich, now Millipore Systems, St. Louis, Missouri.

retracted the skin, and made a similar incision through the abdominal wall (rectus abdominus muscle and peritoneal lining). We then activated a transmitter, made sure it worked, and confirmed its frequency. We used 3.8-gram PD-2T, temperature-sensitive transmitters manufactured by Holohil Systems Ltd. Transmitter batteries were designed to last six months.

After verifying transmitter function and frequency, we sterilized the transmitter in 100% ethanol, dried it, and inserted it through the abdominal wall incision. Using dissolvable sutures, we first stitched up the muscle wall, and then closed the skin using an interrupted pattern of stitching. We applied tissue glue over the sutures to waterproof the skin incision until it healed. We then placed the frog in a clean water bath, nostrils out of the water, until it woke up. We again checked for transmitter function. Once the frog was able to sit upright on its own, we transferred it to the small cooler lined with moist paper towels that we used to transport it from the field. We allowed the frog to recover in the coolness of a refrigerator. The following morning we returned the frog to HFWA-W, where we released it into a clump of dense vegetation outside the drift-fence where we had captured it. Their incisions healed within a few days.

There were a few surgical complications. The most common was from bleeding. While most midline incisions never bled, or bled lightly, some bled so badly that we stopped thinking about implanting transmitters and instead turned our focus to stopping the bleeding. In our estimation, if the animal had bled too much, we stopped the surgery, sewed up the incision, allowed the frog to revive, and took it back to the field the next morning.

The second surgical complication involved frogs recovering from the anesthetic. Postoperatively, we were always anxious until the frog began breathing again. Sometimes we would lightly manipulate the frog—rolling our thumbs along its flank—to stimulate it. This technique failed only once—we had one frog that never woke up from its anesthetic. For us, this loss was a tragedy.* But this was a singular event; Jen successfully performed 61 of 68 transmitter surgeries. Of the seven unsuccessful surgeries, six were stopped

---

* We made another mistake. To assess Crawfish Frog use of wetland habitats during breeding, we implanted transmitters in five males total in Nate's, Cattail, and Big Ponds. While no scientific literature had suggested that implanting transmitters into breeding male frogs was a bad idea, we soon realized that the energy these males use to produce calls that carry over a mile is enough to tear abdominal stitches. Jen collected these males and we discovered all of them had viscera herniated through their anterior abdominal wall. We surgically repaired each of these animals. One of them died during surgery; a second animal died two and a half weeks later. Five months post-surgery, two of these animals (we removed the radio from one animal and its fate was unknown to us) were alive. One of these animals—Frog 060—lived another year-and-a-half (he bred in Nate's Pond in 2011, the last time we saw him). These are tough animals. Details of this regrettable exercise are in Heemeyer et al. (2010a).

because of bleeding, and the animals recovered; the seventh was the frog that never revived.

The following is an example of Jen's surgical notes from 1 April, 2009:

Frog 6 ♂ Big Pond	
422 ml PBS	
0.1 g MS-222	12:17 (pH 7.1)
0.1 g MS-222 added	12:37 (pH 6.95)
Anesthetized	12:57
Surgery completed	13:15
Awakened	13:57
Frequency 150.139	

This was our introduction to Frog 139, who would become one of our rock stars. Note that 139 needed two doses of MS-222 to become anesthetized. The surgery, including suturing, lasted fewer than 20 minutes; it then took this frog almost 45 minutes afterward to recover. This was a typical surgery, with no serious complications. As the notes on 139 suggest, we found Crawfish Frogs to be slow to react to the anesthetic and slow to recover from it. In contrast, we found Crawfish Frogs quick to heal.

Jen would subsequently re-implant 15 of her telemetered frogs with new transmitters after the batteries in their original transmitters gave out; she went on to implant two frogs, including Frog 26 (nicknamed Romeo), three times. On 27 June, 2010, Jen wrote this about switching out transmitters on Frog 160:

> Old transmitter was encapsulated by 2 thin, clear membranes that had blood vessels running through them. Some jelly-like goo [between] the membrane layers. I cut through the clear parts, around the blood vessels and got the transmitter out, membranes were very stretchy.

One frog, 080, shed his transmitter through the side of his abdominal wall. As soon as we realized this, we captured and examined him, expecting to have to sew him up, but the wound had completely healed.

As mentioned above, we experimented with attaching transmitters to frogs using beaded-chain belt harnesses. We attached transmitters this way to 18 frogs, but never felt comfortable for the frogs, which not only had to move through dense prairie grass with a belt trailing a transmitter and an antenna, but also had to turn themselves around in burrows that approximated their girth with this clumsy apparatus hanging off them. We soon removed these transmitters and gave up on the idea, although we would later use belt harnesses to attach transmitters to juveniles—with these tiny frogs we had no other option (Lannoo et al. 2017), and they occupy burrows much larger than their body.

Jen recorded the locations of her telemetered frogs every day.* At each frog's location, she used her Kestrel® portable weather system to record ambient temperature, wind speed, and relative humidity. She also calculated and recorded the body temperature of each frog from signal frequencies sent by their internal, temperature-sensitive radio-transmitters. Jen entered all data into her PDA (personal digital assistant), which she downloaded daily to a hard drive in her apartment, and copied to a hard drive in our lab once a week.

Near each occupied burrow entrance, Jen also placed iButton temperature sensors (iButtonLink, LLC, Whitewater, Wisconsin). iButtons record automatically, and periodically Jen would retrieve them and download their stored temperature data. Using this technology, she was able to compare internal frog temperatures (from their temperature-sensitive transmitters) with local ambient temperatures (from the associated iButtons). She did this for 18 months, from the spring of 2009 to the fall of 2010.

Jen tracked a total of 48 adult Crawfish Frogs, which represented a third to a quarter of all the known Crawfish Frogs at our study site at that time (Heemeyer et al. 2012). She collected data across 7,898 telemetered frog days, the serial equivalent of about 21.5 years. She followed one frog (Romeo) for a total of 606 days—20 months—using three different transmitters. She tracked frogs up to 1.2 kilometers from their breeding wetlands (Figure 4.2).

Jen quickly came to understand Crawfish Frogs, and realized early on that she had to divide occupied burrows into those used during the non-breeding period, which she termed primary burrows, and those used during pre- and post-breeding migrations, which she termed secondary burrows. Jen found that primary burrows were, without exception, crayfish burrows. Her results were unambiguous, and confirmed the early observations of Hurter, Thompson, and Gloyd. Further, Jen found that in the well-drained soils characteristic of our study site, most frogs occupied one, and only one, crayfish burrow for the entire length of time between spring breeding events, an interlude that averaged ~10.5 months and included the late spring, summer, fall, and winter seasons. Gloyd anticipated this result when he wrote: "the fact that it remains quiet throughout the day and in or near its burrow all the year except during the breeding season, may account for its being so little known" (Gloyd 1928).

You might ask yourself, how difficult could this telemetry study have been when Crawfish Frog adults stayed in the same place for 80 to 90% of the year? But we didn't know this at first, and occupied burrows could be

---

* Using a Yagi© unidirectional antenna and an R-1000 receiver (Communication Specialists, Orange, California).

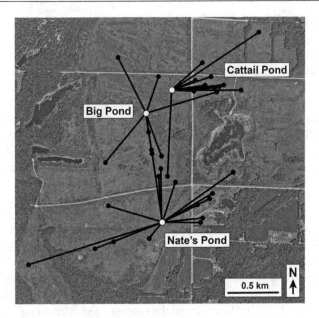

FIGURE 4.2 Locations of Jen Heemeyer's telemetered Crawfish Frogs. Black lines radiating from wetlands indicate frogs captured following breeding at those wetlands and the position of their primary burrows.

over a kilometer from breeding wetlands. Here's how the first field season played out for Jen. She had performed a large number of surgeries over a short period of time, and after she had returned telemetered frogs to HFWA-W they remained near their release site until a heavy nighttime rain occurred; then they all took off. On 20 April, 2009, MJL wrote in his field notes: "All animals moving. Jen overwhelmed trying to find them. Will get her help." The next day MJL followed up:

> Jen's animals moving a lot. Too much to keep up with [she had been out tracking for 15 straight hours]. In a way this is exciting. Animals don't just move away from wetlands and set up shop for summer/fall/winter then return [in fact, yes they do]. Instead, with rainy periods immediately following breeding they move, and sometimes they move a lot. One animal ~0.5 km N of Cattail.

It didn't take Jen long, however, to figure out that when she lost track of a frog, all she had to do was simply keep moving in the direction the frog had been going—along the vector it was taking. It might take some time, and she might walk a long way, but eventually she would pick up its signal and home in on its location. Jen's data showed that a Crawfish Frog adult can move 400–500 meters a night under warm, rainy conditions; a distance over ten times farther than the range of her telemetry equipment.

Jen found that during these post-breeding migrations, Crawfish Frogs used a variety of retreat sites depending on what was available, including crayfish and mammal burrows, prairie grass clumps, and shallow scrapes that frogs dug themselves (this was a surprise*). Authors who claimed Crawfish Frogs use retreat sites such as these were probably observing frogs during their pre- and post-breeding migrations (when Crawfish Frogs are most easily observed). The use of these alternate retreat site types does nothing to diminish the fact that for 10.5 months of the year, when they are not preoccupied with breeding, Crawfish Frogs are obligate crayfish burrow dwellers.†

Crawfish Frogs will occasionally abandon their primary burrows. Jen tracked ten frogs that left their burrows for various periods of time, lasting from days to weeks, before returning. She termed this behavior "ranging." These were usually isolated events, perhaps motivated by the need to seek new, unsoiled burrows. In 1915, Crystal Thompson wrote:

> At the bottom of the frog burrows, which usually terminated at a distance of about three feet, was a mass of foul smelling clayey material containing quantities of beetle remains and considerable dead grass, the latter probably having been washed in or accidentally carried down by the frogs. [According to Roger Thoma and Brian Armitage (2008), crayfish will also incidentally carry in debris.]

As important as crayfish burrows are to Crawfish Frogs, Crawfish Frogs are unencumbered by the responsibility of keeping them clean. When a Crawfish Frog is in its burrow, looking up and facing the opening (as they almost

---

* Vanessa Kinney Terrell tells us Gopher Frogs will do this, also.
† We can look back on the historical literature and know who among the early naturalists was a careful and cautious observer, and who was trusted. These were not always the same people. For example, Wright and Wright (1949) trusted Thompson (1915, p. 6; "Professor La Rue found the frogs in the mammal burrows along the shores of the temporary ponds, as well as in crayfish holes, but it is probable that they were only temporarily occupying the former during the spawning season for we were unable to discover any mammal burrows, either in the vicinity of ponds or elsewhere inhabited by frogs."). Conant (1958), writing later, did not ("Not restricted to crawfish holes, but often found in those which have lost their chimneys and contain water. Other habitats include mammal burrows, holes in roadside banks, and in storm or drainage sewers."). Conant was not wrong, but by ignoring Thompson he presents a much less nuanced, and therefore less-correct description of the relationship of Crawfish Frogs to their burrows. Similarly, Bailey (1943) was such a respected biologist that his observation of two Crawfish Frog juveniles inhabiting a single burrow (which juveniles will do but adults normally do not) has unfortunately been interpreted as a general phenomenon (e.g., Wright and Wright 1949; Minton 2001).

always do), and the urge hits, the deposit often ends up on the side of their burrow. A Crawfish Frog moving up and down its burrow then presses the more resistant elements of this stool, such as grasshopper legs and beetle elytra, against the burrow wall, creating a diet-driven interior décor. On 22 October, 2010 Jen scoped Frog 279's burrow and wrote: "burrow was straight ... then curved, bug bits on walls."

Stools that manage to plunge or tumble all the way down the burrow and into the chamber water at the bottom decompose, foment bacteria and other microbes, and must contribute to the foul smell that Thompson noted. After a period of time (our data suggest this can be as long as five years), the accumulated waste at the base of the burrow must resemble a port-a-potty tank during a county fair, and even the most septic-tolerant frogs are motivated to leave. Again, our best guess is that the ranging movements Jen observed represent failed attempts to find suitable new burrows.

We documented one synchronized burrow exodus. This event occurred on 10 October, 2009, after torrential rain flooded many areas of HFWA-W. Three frogs—two males, 139 and 26 (Romeo), and a female, 220—abandoned their inundated primary burrows to seek shelter in drier burrows. At the time we thought it was strange that ranid frogs would abandon burrows simply because they were wet; ranid frogs are supposed to be all about wet. Later that winter we discovered why. On 10 March, 2010, Jen found Frog 9 dead in his burrow, which had filled to the surface with water and frozen over. During Indiana winters, big rains frequently precede cold fronts, and that's what had happened here. Rain first flooded Frog 9's burrow, and then a cold snap froze it over before the water could percolate out and create an air pocket. Submerged, and without access to atmospheric oxygen, Frog 9 drowned (Heemeyer and Lannoo 2011).

These were the only two instances where we observed Crawfish Frogs affected by flooded burrows. Frogs that inhabit low-lying habitats that frequently flood, for example at Big Oaks National Wildlife Refuge (Robb, pers. comm.), appear to shift burrows more often than frogs in regions such as HFWA-W with better-drained soils. Presumably, this behavior serves to avert Frog 9's fate.

Given the body size of Crawfish Frogs, the largest radio-transmitters Jen could insert had a battery life estimated to be six months (ideally, transmitters should weigh less than 5% of an animal's body weight, and we adhered to this guideline). Given the quality of Holohil, the company that made our transmitters, we were not surprised that most of our transmitters exceeded their nominal lifespan. Even so, there came a time in the fall when the transmitters we implanted the previous spring ran out of power. Jen wanted to track her animals over the winter and during their breeding migrations the following spring, so we had to devise a way to capture Crawfish Frogs in their primary burrows. The literature was no help, since most early workers were interested in collecting frogs, and didn't care about burrow integrity (i.e., they dug them out). But by the fall of 2009 we were becoming aware of the importance of burrow integrity to Crawfish Frog survival, and we needed to develop a method of capturing frogs without destroying their burrows.

We considered several options, including Norrocky pipe-traps (Norrocky 1984), plungers (Simon 2001), and burrowing crayfish mist nets (Welch and Eversole 2006). We also tried securing the collapsible mesh funnel-traps we were using in, wetlands over the burrow entrance. None of these techniques worked, and so we re-thought the problem. We knew that Crawfish Frogs were among the most terrestrial of North American ranids, and decided to try to take advantage of this fact.

We began by flooding occupied burrows. We simply poured lake water from a 5-gallon jerry can into the burrow, then waited for the frog to surface to gulp air, which usually took around 20 minutes (when we mistakenly flooded more aquatic Southern Leopard Frog-occupied burrows it took at least twice that long for frogs to surface). We kept pouring water into the burrow as necessary to keep it topped up, and would watch the water surface for signs of disturbance, indicating the frog was surfacing. This part of the process worked perfectly. What didn't work so well was capturing the frog during the 1 or 2 seconds it took to air gulp and submerge back into the burrow. When we failed, which we almost always did, it would take another 20 minutes or so of pouring water into the burrow and waiting for the frog to surface before we could attempt and fail to capture it again.

We knew to be successful we would need some trick to capture burrow-dwelling Crawfish Frogs without harming them and without destroying the integrity of their burrows. We needed a device that didn't prevent Crawfish Frogs from reaching the surface to air gulp, but would block them from diving back down their burrows afterward. We hit on an idea that Jen made work—a balloon secured to an aquarium hose secured to a bicycle air pump (Heemeyer and Lannoo 2010). Jen inserted the deflated balloon about 6–8 inches into the burrow, and as the water roiled and the frog started to surface she put her left hand on the extended pump handle. When she saw the frog, she pushed the plunger down, inflating the balloon. With the frog's exit blocked, Jen would grab it with her right hand. The whole technique was clumsy, and it was comical during the handful of times when the balloon popped and water geysered up out of the burrow. It probably didn't work more often than it did, but it worked often enough to allow Jen to re-implant a subset of her frogs with fresh transmitters. It was this technique and these transmitters that allowed her, the following spring, to observe that post-breeding Crawfish Frogs return to the burrows they had previously occupied.

In the spring of 2010, the three frogs that Jen had observed synchronously switch burrows the previous fall (see above) began their pre-breeding migrations in an unusual way—by first returning to the primary burrows they had abandoned. This occurred even though for each frog, their original, primary, burrow was farther away from their breeding wetland than their adopted, overwinter burrow. It is almost always the contrary behavior that gives the most insight, and this observation suggests that primary burrows play a critical role in orienting Crawfish Frogs to their landscape.

That same spring, 2010, using replacement radio-transmitters, following Crawfish Frog breeding, Jen tracked eight of her telemetered frogs, including Frog 160 (above), back to the same primary burrows they had occupied the previous year (Figure 4.3). This result was new, unexpected, and, from a conservation perspective, critically insightful. Hurter and Thompson had each assumed that Crawfish Frogs use a number of crayfish burrows interchangeably as retreat sites from which they ranged to feed and breed (a not unreasonable supposition, both species of Gopher Frogs as well as juvenile Crawfish Frogs will do this). Gloyd got closer to the truth when he suggested Crawfish Frogs used the same burrow throughout the year. But no one had considered the possibility that these frogs occupy the same burrow year after year—that for Crawfish Frogs, primary burrows are "home."

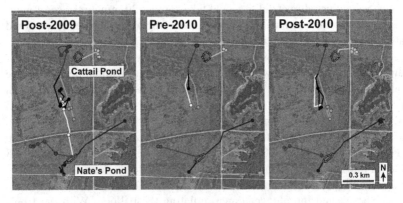

FIGURE 4.3   (Left) 2009 post-breeding movements of Crawfish Frogs from wetland breeding sites to primary burrows; (center) 2010 pre-breeding movements from burrows to breeding wetlands; (right) 2010 post-breeding movements from wetlands back to burrows. Line shadings indicate individual frogs; circles represent sites where frogs were associated with a burrow. Note that in 2010 (right), every frog returned to the primary burrow it had occupied in 2009 (left), even if it had bred in a different wetland (white line). This figure is partially reproduced from Heemeyer and Lannoo (2012) and used with the permission of Leo Smith and the American Society of Ichthyologists and Herpetologists.

After Jen graduated with her master's degree in 2011, Nate and RMS continued to track Crawfish Frog burrow habitation, and by 2014, three frogs—26 (Romeo), 53 (Corner Burn Chick), and 65 (Private)—had each occupied their respective burrows for five consecutive years. When we terminated our study in 2016, Corner Burn Chick was still alive, occupying a new burrow unknown to us. She had bred in every one of the eight years of our study. By 2013 or 2014 we had learned to recognize her big, scarred body. Although as scientists we are not allowed to say this, as human beings we will admit it was always special when we encountered one of these old-timers. Young frogs grow by adding length, old frogs grow by adding girth; and as they go through life, all frogs accumulate scars. After a few years' experience we learned to recognize most of the ancient, beat-up frogs in this population, and with this familiarity came admiration and deep respect.

From her data, Jen estimated home range sizes of individual Crawfish Frogs at around 0.05 m²—about half the size of a footprint. Crawfish Frogs do not put a lot of range into the term home range. In contrast, the average distance of their primary burrows from their breeding wetlands was about 350 m, while the maximum distance was over 1 km (six-tenths of a mile). Crawfish Frogs are either/or animals—either they're spending their time at their primary burrows being highly sedentary, or they're in the process of moving great distances over short periods.

We immersed ourselves in the lives of Crawfish Frogs—we reviewed our techniques and discussed our data all the time. These conversations usually led us to the big-picture implications of the information we were gathering. Because Crawfish Frogs are obligate crayfish burrow dwellers, they do not get to choose where they live; instead, crayfish chose these spots. Crawfish Frogs do not live where upland crayfish are absent, for example, in the too-hard, rocky substrate of the Ozark Highlands, or the too-soft sandy substrate of the Kankakee Sands region of northwestern Indiana. And in areas where burrowing crayfish are present, the only options for where Crawfish Frogs can live come from where crayfish have already dug and abandoned burrows. It's a narrow set of options and a curious phenomenon to ponder. We realized early on that the advantages of abandoned crayfish burrow habitation must outweigh the narrow options afforded by solely occupying such a limited habitat type—that crayfish burrows must somehow be something special.

As we struggled with trying to understand Crawfish Frog crayfish burrow use, we came up with what in retrospect were some wild scenarios. MJL wrote one of our favorites in his field notebook on 23 July, 2009:

> Why do Crawfish Frogs use the burrows they use? i.e. why are some [occupied burrows] 10 m from breeding wetlands while others are +1 km? Perhaps breeding wetlands [are] the wrong way to look at this. What about,

instead, that juveniles find good feeding sites, then find a burrow in the vicinity, then [two years later when they are sexually mature] consider which wetland to breed?

It was much funnier discussing this scenario among ourselves. The conversation usually like this:

Imagine the 'Oh shit' moment as a Crawfish Frog, having chosen as a juvenile a fine burrow based on food availability, two years later, after overwintering, emerges from this burrow now thinking only about breeding, and realizing that the nearest suitable wetland is over a kilometer away.

After our first field season was over, in early November, we sat around and asked ourselves more serious questions about Crawfish Frog crayfish burrow use. We came up with the following list.

1) Are burrows chosen based on vegetation characteristics?
2) Are burrows chosen based on site (slope/aspect) characteristics?
3) Are burrows chosen based on burrow type? Primary burrows vs. secondary burrows.
4) How many crayfish burrows on the landscape are large enough to support Crawfish Frogs?
5) Do Crawfish Frogs stay in one burrow, use sequential burrows, or alternate burrows?
6) Burrow soil type—friable, clay?
7) Nocturnal, diurnal?
8) How do they use burrows?
   • Defense?
   • Rest?
   • Digestion?
   • Hydration?
   • Thermoregulation?
9) Why do they leave burrows?
   • Feed?
   • Thermoregulation?
10) Do they use the same burrows season after season?
11) How do they find burrows to use?
12) How do they find breeding wetlands from burrows?
13) How do they find burrows from breeding wetlands?
14) Can they return to burrows when displaced?
15) Juvenile burrows—where/when/how many/when do they find [their] primary burrow?

16) Why do they abandon primary burrows?
* Flood?
* Fire?
* Mow?
* Prey availability?
* Prey density?

And, as mentioned above, in 2010, we began to get answers.

4.2 WHY CRAYFISH BURROWS?

Compared with burrows dug by other animals, crayfish burrows have two advantages as habitat for frogs. First, upland crayfish typically dig vertical shafts (they can move up to 23 kilograms [50 pounds] of soil per night) until they reach the water table, then excavate a chamber, which becomes a subterranean lakelet—a source of water. Second, crayfish burrows are deep (~1 m), allowing frogs to quickly and easily descend to warmer soil layers below the frost line, a benefit during the winter (in Jen's field notes from 22 October, 2010, we read: "Romeo's burrow 2'7" moderate slope. 2" before [burrow] turns. Could see him move. Could see his back but couldn't see his head"). Frogs occupying crayfish burrows thus have a ready-made resistance to both summer desiccation and winter frost (Figure 4.4). Thoma and Armitage note: "crayfish burrows are insulated from the surface environment, have a relatively constant temperature and humidity, and usually have a constant pool of water at the bottom" (Thoma and Armitage 2008; Thompson also observed this). For Crawfish Frogs, life in a crayfish burrow means immediate and consistent access to summer rehydration and winter warmth.

The question then arises, how many abandoned crayfish burrows are there on the landscape—do they occur frequently enough to be a dependable habitat type for Crawfish Frogs? Following a hot August 2011 prescribed burn at HFWA-W, we conducted a burrow survey over the severely scorched landscape and counted 5,904 burrows in roughly 100 acres. If burrow density at our survey site represented HFWA-W as a whole (~1,800 acres; 18 times the acreage of our burrow survey), this suggests the presence of over 100,000 crayfish burrows at HFWA-W. In their survey of Indiana's burrowing crayfish, Thoma and Armitage found that roughly 40% of crayfish-dug burrows had been abandoned and were unoccupied. From these numbers, there might be as many as 40,000 vacant burrows available to Crawfish Frogs at our study site. If true, this estimate suggests there is more than enough habitat to accommodate

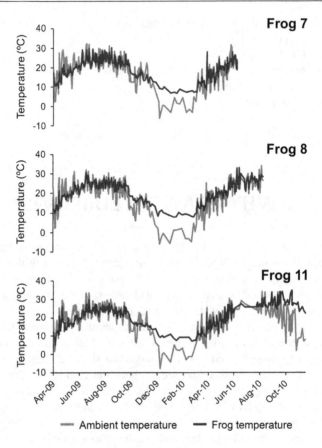

FIGURE 4.4 Air temperatures (gray line, as measured by iButton sensors near the burrow entrance) and frog temperatures (black line, as measured by internal temperature-sensitive radio-transmitters) experienced by three Crawfish Frogs. Measurements were taken continuously for 20 months, from 4 April to 30 November, 2009, at our study site. Note that throughout the warm months, frog and ambient temperatures were similar, as frogs were active and either out of their burrows or near their burrow entrances. During the winter, however, when frogs were quiescent at the base of their burrows, typically a meter or more below the soil surface, frog temperatures were consistently, and often substantially >10°C warmer than air temperatures.

the ~350 Crawfish Frog adults, and perhaps the 5,000–10,000 juveniles that could be produced under optimal conditions in any given year. It appears that at HFWA-W, the availability of abandoned crayfish burrows is not limiting the size of the Crawfish Frog population. This conclusion could change, of course, if declines reported for freshwater crayfish (Edwards et al. 2009) extend to upland species. And, just because our data suggest crayfish burrows are not limiting at HFWA-W doesn't mean they are not limiting Crawfish Frog populations in other regions.

Our burrow survey also indicated that crayfish burrows are not evenly distributed across the landscape; instead they tend to be clustered, as almost everything in natural systems tends to be (Figure 4.5). It also became obvious to us that Crawfish Frog-occupied burrows are included in a subset of burrows that have no chimneys, a large bore, and a bare spot next to the opening. As Thompson indicated, not all burrows that look like this host Crawfish Frogs, but frogs only occur in burrows that have these characteristics. Our data also suggest that burrows in low-lying, frequently flooded areas do not host Crawfish Frogs.

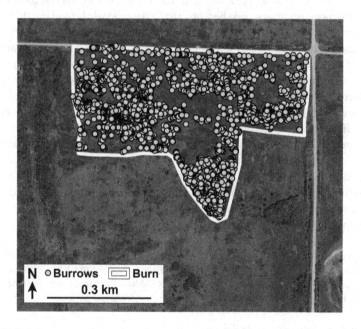

FIGURE 4.5 Approximate locations of crayfish burrows in an area prescribed burned in August of 2011. The fire was so hot that it eliminated all but the woodiest of vegetation cover, and permitted this survey. We located 5,904 burrows. Note that burrows did not occur in wetlands (Big Pond is in the center of the image, Cattail Pond is to its immediate northeast), or in the food plot north of Big Pond and west of Cattail Pond. Note also that burrow locations were clustered, rather than uniformly distributed. Thoma and Armitage (2008) estimate that perhaps 40% of crayfish burrows are abandoned; it's the locations of these vacant burrows that determine where Crawfish Frogs can live during the non-breeding season.

The hydric and thermal advantages of crayfish burrows to Crawfish Frogs are measurable and undeniable. But Crawfish Frogs are thought to have arisen in east Texas (Goin and Netting 1940), an arid land that knows no frost. The couple dozen other frog and toad species that live in this region have found alternative methods to obtain water, and presumably these options would be open to Crawfish Frogs, too. This suggests the burrow advantages Crawfish Frogs possess arose without any pressing need to either avoid summer dry or winter cold conditions. This realization then begs the question, did these hydric and thermal benefits drive the habitation of crayfish burrows by Crawfish Frogs, or do they represent side benefits—epiphenomena—of a behavior driven by some other factor? Over time and after much considered discussion, we have grown to prefer this latter notion. Specifically, we have come to favor the idea that Crawfish Frogs began inhabiting crayfish burrows to avoid predators, and that the hydric and thermal advantages to burrow occupation were side benefits. Here's the rationale behind our thinking.

Crayfish burrow shafts are not only deep, but they are also narrow. While the claustrophobic among us shudder at the thought of being at the bottom of a hole 25 times our height (50 meters) no wider than our bellies, such a situation serves Crawfish Frogs well. Hurter observed Crawfish Frogs on their feeding platform and wrote (in the singular): "As soon as it hears an unusual noise or sees someone it creeps back into the hole." Thompson's description is more dynamic, and in our experience is more correct: "[Crawfish Frogs are] very agile in seeking the concealment of their burrows when alarmed." Our observations (using wildlife cameras, described below) reveal that Crawfish Frogs on feeding platforms nearly always face their burrow entrance, and our video recordings show, and personal experiences confirm, that when frightened they instantly dive into their burrow.

Once in their burrows, Hurter notes that Crawfish Frogs:

were not deeper down than we could reach with our hands … Curiously enough they made no attempt to recede farther when we reached for them. Had they dropped to the bottom we never could have captured them, as some of the holes were three or four feet deep with water at the bottom.

Thompson once again adds detail:

Apparently when alarmed the frogs do not ordinarily descend far into the burrows … When one attempts to dig them out, they descend farther into the holes, the exceptionally powerful hind limbs and the extent to which they can distend the body serving to secure them so firmly that they could be mutilated before being dislodged.

Thompson also noted: "The frogs so nearly approximate the size of the holes that the rubbing of their soft bodies probably tends to smooth the walls."

Phil Smith almost got it right. In his fine book, The *Amphibians and Reptiles of Illinois* (Smith 1961), he writes, "The frog usually may be seen an inch or two below the surface, but when disturbed it backs down to the bottom of the burrow, where it wedges its body against the side of the hole." The two details he implies but does not state are 1) that Crawfish Frogs in their burrows face the entrance and 2) that they inflate to wedge their body against the burrow walls. What is unlikely, in our experience, is that Crawfish Frogs back all the way down their burrow before inflating. Crayfish usually excavate a chamber at the bottom of their burrows (where, using a burrow scope, we have seen Crawfish Frogs sit during the winter), making it impossible for a frog to wedge itself. (It is possible that frogs in the chamber at the base of their burrow inflate to a size larger than their burrow shaft diameter, which would make it impossible for them to be brought to the surface [think Pooh bear and the honey pot]. But in fact, a snake could easily kill a frog inflated in its chamber, which would cause it to deflate [see below], and be removed through the burrow shaft.)

These old-timers described most of the behavioral components of the Crawfish Frog–crayfish burrow relationship, but never quite came to realize how all these parts fit together—along with the shape of the adult Crawfish Frog head—into a superbly effective defensive strategy. It took a tragedy, but Jen and Nate finally began to figure it out (Engbrecht and Heemeyer 2010). From Thompson's observations they knew: 1) Crawfish Frogs "so nearly approximate the size of the holes that the rubbing of their soft bodies probably tends to smooth the walls"; 2) "the exceptionally powerful hind limbs and the extent to which they can distend the body serving to secure them so firmly that they could be mutilated before being dislodged"; and 3) "when alarmed the frogs do not ordinarily descend far into the burrows." Jen and Nate's insight occurred on 14 May, 2009. Nate, tracking Jen's telemetered Frog 520, a female, found the signal coming from outside her burrow and moving faster than a Crawfish Frog could possibly move. Nate caught up to it, pounced, and came up with an Eastern Hognose Snake sporting a suspicious mid-gut bulge. Sure enough, the signal was coming from the bulge. (Jen's data would eventually demonstrate that snakes are major Crawfish Frog predators.) When Nate grabbed the snake, it—being guilty only of trying to make a living—got frightened and simultaneously defecated and vomited. And up came 520's carcass (Figure 4.6).

As mentioned above, it's the things that don't fit that often prove most insightful, and when Nate examined the Crawfish Frog carcass what didn't fit was that this frog was missing its head. The head hadn't fallen off and could not have been completely digested overnight. Instead, it was obvious from the slashed and torn tissues in the neck that 520's head had been ripped from her body (mutilated, in Thompson's phrasing). Except for this, the carcass was undamaged. Unlike sharks and crocodiles, snakes do not tear apart or chew

FIGURE 4.6   Frog 520's body the morning after she had been eaten by a Hognose Snake. Note that her head has been ripped off, and she has no punctures or lacerations on her hind legs. This pattern of trauma suggests she was in her burrow in a defensive posture (inflated, head down, facing the entrance) and attacked head-on. Our best guess is that as 520 was being attacked, she lifted her head, the snake got purchase on her snout, began violently tugging on the inflated and wedged frog, and eventually ripped off her head. The decapitated frog then deflated, and the snake ingested her body. This photo was originally published in Engbrecht and Heemeyer (2010) and is used with permission of Robert Hansen and the Society for the Study of Amphibians and Reptiles.

their prey prior to swallowing it, so what could have happened here? MJL remembers clearly where he and Engbrecht stood talking with Heemeyer on the phone, near his truck, parked at the trailhead to Nate's Pond, as they worked this out. Since this frog was attacked head-on, it had to have been in its burrow. (A frog attacked on land without the benefit of such a refuge will flee, and will therefore be attacked at the rear of its body, usually bitten on and secured by its hind legs. Frog 520 had no hind-limb punctures or lacerations.) Our best guess was that as 520 was in her burrow, facing the entrance, as Crawfish Frogs always do, and about to be attacked, she lowered her head, puffed up her body (Figure 4.7) (see also Altig 1972), wedged herself against her burrow wall, then put up a great resistance, as Thompson described. Subsequently, 520 made a mistake—she must have lifted her head. The snake got purchase on her snout, began violently tugging on the inflated and wedged frog, and eventually ripped off her head (again, the "mutilation" Thompson predicted). The decapitated frog then deflated, the snake ingested its body, and a few hours later Nate showed up.

FIGURE 4.7   A Crawfish Frog male in a defensive posture with its body inflated. Now, imagine this frog is in its burrow, facing the entrance. When he lowers his head, he allows his bony rounded snout to act as a shield separating the softer portions of its body from a snake predator.

The fate of these snake–Crawfish Frog encounters do not always, and may not usually, favor the snake. Using wildlife cameras to follow the behavior of Crawfish Frogs, we have been able to document several snake-on-Crawfish Frog predation attempts. One stands out. On 4 July, 2011 (Independence Day, in at least two senses of the word, as we are about to see), at 2:00 PM, we recorded Frog 26 (Romeo) on his feeding platform facing his burrow entrance (Figure 4.8). An hour later (the finest time resolution of the early-model wildlife cameras we used) Romeo was gone and a Black Racer had its head inside Romeo's burrow, working it hard. An hour later, and for the next three hours after, neither animal was visible. By 8:00 PM, Romeo had emerged and assumed the exact same position and orientation that he had shown six hours earlier—as if nothing had happened—with no blood and no scars, appearing ready to eat the next insect that happened by.

To reiterate, what we now know—and what the old-timers did not, even though they had assembled most of the behavioral and natural history pieces— is that when on their feeding platform, Crawfish Frogs almost always face their burrow entrance, prepared to quickly dive into their burrow when threatened by a predator. Once in their burrow, they descend only a short way (for example, Thompson 1915), rapidly turn around to face the threat, inflate their body against their narrow and perhaps fitted burrow walls, and lower their head. Against snakes—the predator most likely to enter a burrow—this behavior usually works. It does not work, however, when a frog lifts its head (as we suspect

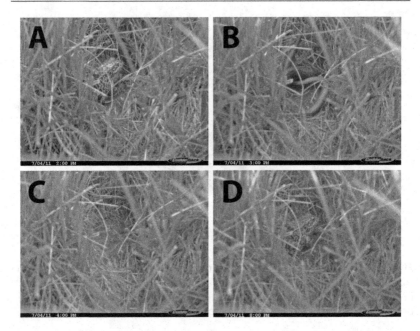

FIGURE 4.8 The details of a Black Racer attack on Romeo, which occurred on 4 July, 2011—Independence Day—as recorded by a wildlife camera set to take pictures at one-hour intervals. A) At 2:00 PM Romeo is outside his burrow, facing the burrow entrance, as Crawfish Frogs usually do. B) At 3:00 PM a Black Racer is attacking Romeo, and experienced snake biologists tell me this snake is aggressively working this burrow. C) At 4:00 PM, neither snake nor Romeo is present, and this would be the case for the next three hours. D) At 8:00 PM, Romeo is again out on his feeding platform, facing his burrow entrance, in almost exactly the same position and posture he was assuming prior to the snake attack. Note that Romeo is intact, with no bleeding or scarring suggesting wounds. We know of no other midwestern frog species that could survive such an attack. This photo series was originally published in Engbrecht et al. (2012) and is used with the permission of Robert Hansen and the Society for the Study of Amphibians and Reptiles.

520 did in response to the Hognose Snake), occupies a burrow too wide to use as leverage, or continues to flee down its burrow. Hurter chastised Crawfish Frogs for not descending farther down their burrows, since this meant they could not elude his grasp. But Crawfish Frogs did not evolve their burrow defense to avoid capture by human hands; snakes are their biggest threat and thus create the greatest selective force in molding their defensive strategy.

(We have also wondered whether toxins might assist Crawfish Frogs in their defense. Crawfish Frogs have numerous large glands on their back that secrete toxins, but these toxins are thought to have antimicrobial rather than anti-predator functions (Ali et al. 2002)).

In essence, Crawfish Frogs use crayfish burrows the same way turtles use their shells. And like a turtle shell, for this behavior to be successful, a crayfish burrow must fit the body of its Crawfish Frog occupant. As Thompson pointed out, as frogs inhabit burrows, the repeated entering and exiting (especially during wet weather when the sides are muddy) causes the burrow opening to conform to the cross-section of the frog, producing a burrow that eventually fits the frog. This may explain why Crawfish Frogs exhibit such intense site fidelity to their primary burrows, both within and across years. Finding a burrow is risky—not only are Crawfish Frogs outside their burrows vulnerable, but a novel burrow may contain an animal that will harm a frog (see below). Further, fitting a frog to its burrow takes time, and during this process the frog is vulnerable. Better for a frog to stay put, even if it stinks up the place.

The relationship adult Crawfish Frogs have with their burrows represents a true adaptation, a novel behavioral trait invented by them that greatly increases their chances for survival. Neither of the Gopher Frog species appears to exhibit this behavior (the use of wide-mouthed Gopher Tortoise burrows precludes this), nor do juvenile Crawfish Frogs, which occupy burrows similar in size to those used by adults, and are therefore much larger than they are. However, for adult Crawfish Frogs, this tactic represents a beautifully effective adaptation, one that allows them to thwart a direct attack by a snake. No other North American frog has this sort of protection. Once adult Crawfish Frogs conjured this idea, we believe they used the secondary benefits of burrows—hydration and thermal inertia—to expand their range into arid and cold regions. Dinosaurs invented feathers for thermal purposes; they came in handy later when a group of their descendants we call birds invented their version of flight.

This defensive strategy doesn't always work for Crawfish Frogs on modern landscapes. We'll provide details below, but offer a hint here. Again, Thompson: "Apparently when alarmed the frogs do not ordinarily descend far into the burrows, for they are plowed out in numbers and the ground in that region is only plowed to a depth of about three inches."

We have recently come to realize that the shape of the adult Crawfish Frog head may assist in their specialized defensive behavior. That is, not only have adult Crawfish Frogs invented a novel suite of behaviors associated with using burrows to resist predatory attacks, but they also appear to have modified the shape of their heads to facilitate their defense. We explore this topic in the next chapter.

Put a Lid on It

5

The study of animal body size and shape is called morphometrics. As a discipline, morphometrics has largely gone out of style, replaced by more technical subjects such as genetics. But an animal interacts directly with its environment through its body, not its genes, and the size and shape of an animal remain relevant factors for field-based ecologists to consider. A biologist who knows what questions to ask can still get a spreadsheet full of meaningful morphological data using a 25-cent ruler.

The first thing to understand about morphometrics is that size and shape are often linked. Across species, size helps determine shape. As Stephen Jay Gould often pointed out, if a mouse was as big as an elephant, it would look like an elephant (because massive bodies need thick legs). Within species, shape often varies with size (growth) in predictable ways. For example, among most vertebrates, early embryonic growth favors the head, while later post-embryonic growth favors the trunk and limbs. Take a photograph of a 2-year-old and blow it up so the child appears adult-sized; their head is the size of a beach ball. Take a photograph of an adult and shrink it to the size of a 2-year-old; they become a pinhead.

The study of differential growth is called allometry (different measure). The reason beach ball heads don't happen to living people is that as human bodies grow, head size grows proportionally slower than the body as a whole, a term called negative allometry, while the trunk and limbs grow faster than the body as a whole, called positive allometry. Again, this is the general pattern among vertebrates, but there are exceptions. For example, freshwater sunfish typically grow proportionally throughout their life. Take a picture of a juvenile Bluegill, blow it up to the size of an adult, and the only difference you'll notice is in the size of the eye (proportionally larger in the juvenile). This pattern of evenly distributed growth is called isometry (same growth).

Allometry has been explained as follows. In newly conceived embryos the zone of most intense growth is located at the snout, producing a large head. As the animal ages, this growth zone migrates toward the tail, producing a faster-growing trunk and limbs, until it eventually ends near the pelvic area as animals reach maturity.

Within vertebrate lineages, this pattern of allometric growth is generally conserved. But it doesn't have to be, and evolutionary novelties often arise through deviations in allometric growth patterns (i.e., through a change in the *timing* of gene expression, not through *mutations* of the genes being expressed, a term called heterochrony). The area of science that explores how developmental shifts drive evolutionary change is informally called Evo Devo.

One of the fundamental concepts of Evo Devo is that larger differences between juvenile and adult shapes produce a greater potential for evolution to create novelties through heterochronic changes. This sounds like a mouthful, but as Pere Alberch has pointed out, it's easy to visualize using human-driven selection for dog and cat breeds (Alberch 1986). A newborn puppy with its pudgy snout, sausage body, and stumpy legs looks very little like most adult dogs. Kittens, on the other hand, look a lot like adult cats. The next time you're in your veterinarian's office, look at the universally present poster of dog breeds, and compare the size and shape of a Pekingese to St. Bernard. Now look at the corresponding poster of cat breeds. You'll see long-haired and short-haired breeds, and the strange color patterns of Calicos and Siamese, but you'll see nothing in the range of size and shapes of cats that resembles the variation observed among dogs. Why? Because the relatively small difference in allometric change between kittens and cats gives breeders much less variation to work on compared with dogs—there is less heterochronic potential.

We were interested to know whether Crawfish Frogs had any specializations in body size or shape that better enabled them to inhabit crayfish burrows. We knew that to do this, we couldn't simply look at adult morphology. That instead we would have to compare growth patterns (allometric trajectories) within Crawfish Frogs and across the three other species in the subgenus *Nenirana* in order to explore possible heterochronic shifts. So we did.*

As we worked through the literature on burrowing frogs, we came to understand that while each of the *Nenirana* species inhabits retreats, none are truly fossorial in Hildebrand's sense of an animal adapted to effectively dig (Hildebrand 1974). Crawfish Frogs and Gopher Frogs will excavate shallow retreats by digging backward (Parris 1998; Humphries, pers. comm.), but

---

* We examined the three other *Nenirana* species to provide an appropriate outgroup comparison. Our measurements were similar to those taken on this species group by Goin and Netting (1940) and on Crawfish Frogs by Bragg (1953), but while these authors were examining characters useful for distinguishing species, our study examines the morphological correlates of burrow habitation. Therefore, we differ from these studies in two important ways: 1) in the use of Pickerel Frogs for outgroup comparison (to provide directionality to across-species differences), and 2) we examined not only adults but also a range of juveniles to provide a post-metamorphic ontogenetic series. Additionally, we analyzed a larger and more geographically widespread dataset than the one available to these authors. For descriptions of the usefulness of ontogenetic trajectories as phylogenetic character states see Mabee (2000) and Maglia et al. (2001).

none of the *Nenirana* species have hardened spades on their feet (similar to Spadefoot Toads) or are known to have limb specializations (shortening, thickening) associated with burrow excavating. Rather than dig their own burrows, members of the *Nenirana* generally inhabit natural cavities or burrows dug by other species.

To compare cranial and hind-limb morphology across all four *Nenirana* species, we obtained Crawfish Frog, Gopher Frog, Dusky Gopher Frog, and Pickerel Frog specimens from museum collections around the country.* We examined 161 animals, including Crawfish Frogs—48 males, 17 females, and 8 recently metamorphosed juveniles (<35 mm snout-urostyle length [SUL], see below); Gopher Frogs—ten males, 9 females, 3 adults whose sex could not be definitively determined, 6 older juveniles (between 50 and 70 mm SUL), and 3 recently metamorphosed juveniles; Dusky Gopher Frogs—1 male, 13 females, 1 older juvenile, and 2 recently metamorphosed juveniles; and Pickerel Frogs—27 adults, 9 older juveniles (between 35 and 40 mm), and 4 recently metamorphosed juveniles. We recognized that juveniles were relatively underrepresented in this sample, but they are underrepresented in most museum collections. Crawfish Frogs and Gopher Frogs are so secretive that they are generally only encountered, and therefore collected, as noisy breeding adults.

Preserved museum specimens are subject to physical shrinking and expanding after being immersed for years in a standard 70% ethanol solution. Since bones shrink less than soft tissues, we decided to radiograph (x-ray) our specimens, and take our measurements from these films. A burrow-based lifestyle is known to influence body size, head size and shape, and hind-limb length. So we measured overall body length, which in radiographs of frogs is the snout-urostyle length (SUL), head width (HW), and head length (HL), as well as thigh and lower leg lengths (femur [FEM] and tibiofibula [TIBFIB] lengths, respectively). Although we were curious about forelimb lengths, we

---

* For details, see Engbrecht et al. (2011). The following scientists loaned us specimens: R. Drewes, J. Hanken, T. Hibbitts, K. Krysko, J. Losos, D. Lowe, M. Nickerson, R. Nussbaum, A. Resetar, S. Rogers, G. Schneider, R. Stoelting, and R. Williams. Museum numbers of specimens we examined are as follows: CAS-SU 2174–80; CM 5407, 13371–75, 13378, 18184–97, 69961, 69962; FMNH 121690, 21741, 21743, 26417, 48217–20, 48222, 48227, 94321, 94323; UF 26, 2375-6, 35370–73, 64262, 66650–53, 87142, 99855–58, 103332, 103333, 111130, 111132; MCZ 7043, 7044; INSM 24, 25; ISU 2, 395–97, 399, 400, 449–52, 818, 937, 966, 1009, 1492, 1522, 1820, 1865, 2255, 2333, 2473, 2738, 2739-Ra 7, 2739-Ra 9, 2739-Ra 10, 2739-Ra 11, 2739-Ra 12, 2739-Ra 14, 2739-Ra 15, 2783, 2822, 3204–07, 3248-Ra1, 3248-Ra2, 3644, 3665; PU 8482, 8483; TCWC 66467; UMMZ 100304, 101623, 103361-1654, 103361-1655, 105544-1984, 108125-2337, 108125-2336, 110638-2568, 110638-2569, 118078. We also examined seven additional specimens subsequently deposited in the collection of the Field Museum of Natural History. (FMNH 288547-288590; Appendix III). Museum abbreviations are listed at http://www.asih.org/codons.pdf.

did not measure forelimb elements because in most museum specimens, fore-limbs are positioned in an up-down (dorsal-ventral) orientation, which makes measuring bone lengths from radiographs next to impossible without damaging the specimen.

In order to compare body proportions, we divided our HW, HL, FEM, and TIBFIB measurements by SUL to obtain ratios of body parts to the animal as a whole. For our analysis we assumed that Pickerel Frogs are basal within the *Nenirana* clade, meaning their morphology approximated the ancestral condition for the group. Further, we assumed that Crawfish Frogs were the most derived species, based on their behavioral specialization of being obligate crayfish burrow dwellers.

Our results showed that the four species of *Nenirana* differ in size. Pickerel Frogs are the smallest species ($\bar{x}$ = 44.9 mm SUL), Dusky Gopher Frogs ($\bar{x}$ = 72.9 mm) and Gopher Frogs ($\bar{x}$ = 79.3 mm) are intermediate in size, and Crawfish Frogs ($\bar{x}$ = 87.2 mm) are largest. This result was not surprising, and our values corresponded to the lengths and relative sizes of these four species reported in the literature. Within species, females tend to be larger, but in our samples sexual differences in length were not significant.

Species within the *Nenirana* also differ in shape. The major distinction is between Pickerel Frogs and the Gopher Frog/Crawfish Frog group. Pickerel Frogs are more slender and have relatively narrower heads and longer legs than either Gopher Frogs or Crawfish Frogs. You don't need a ruler to know that Pickerel Frogs look very different from Gopher and Crawfish Frogs. In fact, this distinct morphological fault line should be sufficient to place Gopher and Crawfish Frogs in their own genus, separate from Pickerel Frogs by the criterion of Bob Inger: "each genus should represent the same kind of entity: a distinct mode of life and a distinct evolutionary shift" (Inger 1958).

Within the Gopher Frog/Crawfish Frog group we could find no significant differences in shape metrics between the two Gopher Frog species, but we did find what may turn out to be an important difference between Crawfish Frogs and Gopher Frogs—Crawfish Frogs have relatively wider, shorter heads than Gopher Frogs; a distinction that sounds trivial, but may not be.

The heterochronic shifts we measured were illuminating. With growth, Pickerel Frogs exhibit strong negative allometry in both head width and length (i.e., as with most vertebrates, Pickerel Frog heads get proportionately smaller as they grow). As with Pickerel Frogs, Gopher Frogs exhibited negative allometry in head length. But in contrast to Pickerel Frogs, Gopher Frog head width remained isometric—proportional to the body as a whole throughout growth. Similar to Pickerel and Gopher Frogs, Crawfish Frogs also exhibited negative allometry in head length. But in contrast to the other *Nenirana* species, and vertebrate tendencies in general, their head width demonstrated a robust positive allometry—with growth their heads got proportionately wider. To reiterate, within the *Nenirana*, head length always exhibits negative allometry, but head width exhibits a directed change from negative allometry in the ancestral-like Pickerel Frogs, through isometry in the intermediate Gopher Frogs, to positive allometry in derived Crawfish Frogs.

	Pickerel Frogs	Gopher Frogs	Dusky Gopher Frogs	Crawfish Frogs
Head Length	↓	↓	↓	↓
Head Width	↓	=	=	↑

A conceptual way to look at this is that during development in the *Nenirana*, the zone of optimal growth migrates from head to tail (i.e., heads grow proportionally faster early, the trunk and limbs grow proportionally faster later), as expected for most vertebrates. In Pickerel Frogs, this zone migrates relatively quickly to present a pattern of negative allometry in both head length and head width. In Gopher Frogs, this growth zone slows down (a heterochronic shift) enough to preserve negative allometry in head length but isometry in head width. In Crawfish Frogs, this growth zone slows down even more, presenting—as with all *Nenirana* species—a negative allometry in head length, but producing a positive allometric change in head width—a reversal of the ancestral pattern. This allometric about-face in Crawfish Frog head width is consistent with the positive allometric pattern observed in the trunk and limbs. One result of this allometric shift is that head shape in juvenile Crawfish Frogs more closely resembles head shape in adult Pickerel Frogs than it does adult Crawfish Frogs (Figure 5.1).

Confirming our measurements, Goin and Netting (1940, p. 146) write, "The shape of the head when viewed from above is orbiculate in [R.] *a. circulosa* [Northern Crawfish Frogs], U-shaped in [R.] *a. areolata* [Southern Crawfish Frogs], triangular in [R.] *sevosa* [Dusky Gopher Frogs], and subtriangular in [R.] *capito* [Carolina Gopher Frogs]." Further, Bragg observed that in Crawfish Frogs "the larger the size, the rounder the snout" (Bragg 1953, p. 280). This snout rounding is due, in fact, to the relative increase in head width relative to head length that we describe above. Positive allometry in head width is the one single, novel heterochronic trait expressed by Crawfish Frogs. Given this, we suggest there may be functional consequences to snout rounding for a burrow-dwelling frog trying to avoid predators. A rounded snout offers the opportunity for Crawfish Frogs to seal a cylindrical burrow with their heads (i.e., a rounded head minimizes the gap between the burrow wall and the jawline of a frog occupying that burrow, providing less opportunity for predators such as snakes to gain purchase on the jaws). In effect, when a Crawfish Frog is in its burrow facing up, toward the opening of the burrow (as they almost

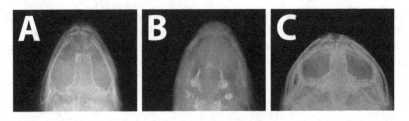

FIGURE 5.1   Crawfish Frog skull shape may contribute to their ability to survive snake attacks, such as the one experienced by Romeo. Here we show radiographs of three skulls as follows: A) a 40-mm SUL Pickerel Frog, skull width 14 mm; B) a 24 mm juvenile Crawfish Frog, skull width 9 mm; and C) a 95 mm adult Crawfish Frog, skull width 38 mm. We adjusted magnifications to present images with similar head lengths in order to better compare head widths and skull shapes. Note the similarity in skull proportions between the juvenile Crawfish Frog and the adult Pickerel Frog, and how rounded the adult Crawfish Frog skull is by comparison. With growth in Crawfish Frogs, skull morphology shifts from the pointed snout of juveniles to the rounded snout of adults due to differences in the timing of growth (heterochronic shifts) in head length and head width. You can ask yourself, if you needed to block a cylindrical burrow shaft, which skull shape would you prefer?

always do), and lowers its head at the approach of a predator, its bony, rounded head acts as a hatch that seals the frog's soft body from a potential predator (Figure 1.3). This morphology only works with an animal that occupies a burrow the diameter of their body.

Are there other reasons for Crawfish Frogs to have a rounded head? From the scientific literature, snout morphology is known to be influenced by at least two functional demands, prey size and headfirst burrowing. As far as prey size is concerned, amphibians are gape-limited predators. That is, amphibians swallow their prey whole, and the size of their gape provides an upper limit to the size of prey they can ingest (Zaret 1980). Dundee and Rossman refer to gape limitation in Crawfish Frogs when they tie the size of their prey (other frogs and insects) to their head width (1989, p. 108). It is not clear, however, that the general diets of Crawfish Frogs differ enough from Gopher Frogs to warrant a divergence in snout morphology. Headfirst burrowing also seems an unlikely explanation of Crawfish Frog snout morphology. Crawfish Frogs do not dig forward, and they do not exhibit the forelimb specializations (Brown et al. 1972) or cervical flexibility of some forward digging species (Emerson 1976; Davies 1984; Nomura et al. 2009). If, indeed, snout rounding represents a morphological contribution to Crawfish Frog predator defense, it would offer a third, novel function for the blunt snouts of burrow-dwelling frogs.

Two loose ends. In our study, hind-limb lengths were too variable to assign much meaning to differences across *Nenirana* species. It is worth noting that among true fossorial frogs, hind-limb diggers such as *Glyphoglossus molossus* (Emerson 1976) have short hind limbs, while Crawfish Frogs, as with all members of the *Nenirana*, have long hind limbs. Both adults and juvenile Crawfish Frogs will, however, scrape the soil to dig shallow retreats when burrows are unavailable. Their legs are not built for this, though, and they do a lousy job of it (see Engbrecht et al. 2011).

We also don't know what to make of the data showing Crawfish Frogs are the largest *Nenirana* species. Earlier workers reported a north–south gradient in size and body proportions in Crawfish Frogs (Goin and Netting 1940; Bragg 1953). Recently, MJL visited Attwater Prairie Chicken National Wildlife Refuge during Crawfish Frog breeding season. He caught a calling male Southern Crawfish Frog that was 72 mm long; if he had seen this mature male in Indiana, based on its size he would have considered it a juvenile. Consistent with this observation, museum specimens suggest that the largest Crawfish Frogs are in the northernmost portions of their range (our largest Crawfish Frogs, gravid females, were 115 mm SVL and 188 grams* and 120 mm SVL and 182 grams†). Given this latitudinal gradient, the appropriate Crawfish Frog size comparison with Gopher Frogs would be Southern Crawfish Frogs in Texas, which, as Goin and Netting have noted (1940, p. 146), are likely no larger, and indeed may be smaller than Gopher Frogs inhabiting the same latitude along the Gulf Coastal Plain.

Back in 2011, when we published this morphological work, we did not fully comprehend how Crawfish Frogs use burrows in predator defense. For example, we did not understand the importance of tailoring burrow to frog, nor had we observed that Crawfish Frogs would occupy the same burrow for half a decade, presumably because it fits them and by fitting them allows them to survive. From this perspective, we feel the heterochronic shift in head width in Crawfish Frogs to produce head rounding is important and may have functional, life-or-death, consequences.

This idea, we feel, remains open to interpretation. We've thought about how to test the adaptive nature of rounded snouts. Perhaps first set up enclosures with Crawfish Frogs in appropriately sized burrows, and then release appropriate (size and species) snakes and determine the fate of the frogs. But we cannot create Crawfish Frogs with pointed snouts and compare survivorship in them to frogs with normal, rounded snouts. We could populate half the burrows with more pointy snouted Gopher Frogs, but Gopher Frogs have not been reported to possess the burrow-based defensive behaviors that Crawfish

---

* Pit tag number 4b08004020.
† Pit tag number 4109543420.

Frogs exhibit (facing the burrow entrance, puffing up their bodies, lowering their heads), so may be disadvantaged in snake attacks not from their morphology, but rather from their behavior. Perhaps a burrow could be half-excavated and a Plexiglas™ wall installed—similar to what nature centers do with bee colonies—to observe snake–Crawfish Frog interactions. We can't imagine the logistics required to pull this off. What we are left with is the difficult-to-test idea—which at this point is just an idea—that Crawfish Frog heads become rounded as they age in order to provide a lid—a closed hatch—that provides an effective seal separating a predator from the body of the frog. It sounds crazy, but hey, these are Crawfish Frogs, so why not?

Far, Far from Me

6

In the early spring, Crawfish Frogs leave their primary burrows to breed in fish-less seasonal wetlands (shallow basins that generally retain water into the summer) and semi-permanent wetlands (deeper basins that remain wet throughout the year except during droughts). Demonstrating the value of burrows in reverse, Jen showed that Crawfish Frogs migrating to and from breeding wetlands were 12 times more likely to be preyed upon than frogs in burrows. This is unfortunate, but the animals themselves must take some responsibility for this carnage. One factor Crawfish Frogs do not appear to consider when choosing their primary burrow is its closeness to their potential breeding wetland. In Jen's study, the average distance of primary burrows from breeding wetlands was about 350 meters (over three football field lengths). The average adult Crawfish Frog is about 10 centimeters long, so its average migration distance is about 3,500 body lengths. For a human, that would be like walking about 7 kilometers (4-and-one-third miles). The most distant burrow Jen found (Frog 22 [780]) was over 1 kilometer from this frog's breeding wetland, the human equivalent of walking about 16 kilometers (10 miles). The urge to breed must be strong.

By now, it should be obvious that Crawfish Frogs challenge whatever generalizations we might make about them based on the biology of other species of frogs, and their breeding habits offer no exception. In the Upper Midwest, among pond-breeding amphibians, small-bodied frogs such as Spring Peepers, Chorus Frogs, and small ranids, including Wood Frogs and Leopard Frogs, breed in the spring and have tadpoles that metamorphose in mid- to late summer. In contrast, large-bodied ranids, such as Bullfrogs and Green Frogs, breed in the early summer and usually have tadpoles that overwinter at least once before metamorphosing. Based on body size alone, one expects Crawfish Frogs to breed late and have tadpoles that overwinter.

Despite this prediction, the earliest historical reports of Crawfish Frog reproduction all pointed to spring breeding: Thompson (1915: March–April), Wright and Myers (1927: March–April), Gloyd (1928: March–mid-April), Wright and Wright (1933: March–April), and Cagle (1942: March–April). Consistent with spring breeding, early authors such as Wright and Wright (1933) observed

Crawfish Frog tadpoles metamorphosing in July and August. This would appear to be a solid foundation for understanding the timing and duration of Crawfish Frog breeding and larval development. But in 1943, Reeve Bailey, normally a reliable observer, inexplicably stated,

> Three early spring collections of overwintering [Crawfish Frog] tadpoles show no sign of approaching metamorphosis, but since unimodal size distribution is evident … transformation may be assumed to occur during the ensuing summer." Bailey, who was careful and cautious, and perhaps suspicious of these data, hedged his bets: "… it is possible, however, that [a small adult male] transformed early in the summer of his first year and was mature at two years of age.
>
> (Bailey 1943)

Breeding must be exhausting for Crawfish Frogs. Before breeding, Crawfish Frogs overwinter at the base of their burrows. They do not hibernate or exhibit any form of torpor (nor do they sleep, a topic we consider in Chapter 15). When Jen inserted burrow scopes down occupied burrows in the dead of winter, she saw awake, alert, attentive Crawfish Frogs looking up at her from the chamber at the base.

In the spring, once the frost has left the soil, Crawfish Frogs emerge from their burrows to feed on early-season invertebrates. At this time, adults range farther from their burrow entrances than they do following breeding. On 12 March, 2013, our field notes read: "Private, Romeo, and Corner Burn Chick (pre-breeding) out of burrows and around burrows more than we ever see after breeding." Also during this time, males call from their burrow entrances. These upland calls (detailed in Chapter 7) differ in pattern and volume from breeding calls (Engbrecht et al. 2015). The function of these pre-breeding calls is unknown, but the effect of these soft songs originating from scattered upland sites across a spring prairie landscape presents an aura, a hint of what is to come, since the first breeding migrations are just a nighttime thunderstorm away. Such an overture seems like cheating; Mother Nature rarely shows her hand.

Once Crawfish Frogs emerge from their burrows in the spring, a hard, warm, nighttime rain initiates breeding migrations. Not all rains trigger migrations in all frogs, and because not all frogs are equidistant from their breeding wetlands, or move at the same rate, or are the same sex, frogs do not show up at their breeding wetlands simultaneously (Figure 6.1). Crawfish Frogs are capable of moving between 400 and 500 meters a night, at a rate of perhaps 60 to 80 meters/hour. If they don't reach their wetlands before dawn they seek shelter or hunker down, sometimes in a makeshift shallow scrape they've dug (see Chapter 5).

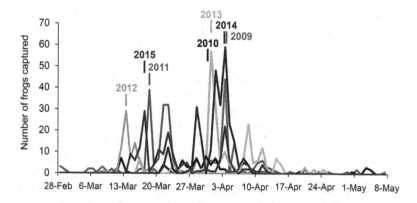

FIGURE 6.1 Crawfish Frog breeding periods based on animals captured at drift-fences at our study site in southwestern Indiana. Note that peak breeding varied from 13 March, 2012, to 4 April, 2009, and 2014. Experienced Hoosier biologists offer that peak Crawfish Frog breeding occurs on St. Patrick's Day, which, indeed, turns out to be more accurate than many of these almanac-type pronouncements. This image was originally published in Lannoo and Stiles (2017) and is used with the permission of Leo Smith and the American Society of Ichthyologists and Herpetologists.

Each year, we became preoccupied with the logistics of documenting Crawfish Frog breeding right after the New Year's celebrations wore off, paradoxically the coldest, snowiest time of year. MJL remembers a blizzard in 2009 at the same time equipment and supplies were rolling in for our first field season. It seemed absurd. Yet, in southern Indiana, the snow melts fast. By the end of February it warms up enough that Wood Frogs* and Smallmouth Salamanders† are moving into breeding wetlands, and a week or two later Chorus Frogs and Spring Peepers start calling. There is an old wive's tale among Hoosier herpetologists that Crawfish Frogs begin to call around St. Patrick's Day. It's not a bad approximation. Populations in the south, near the Ohio River, call a few days to a week earlier than populations in the north (at about the latitude of Indianapolis). In the Midwest, spring moves northward at a rate of roughly 16 kilometers (10 miles) a day.

In 2009, the concerns leading up to our first field season centered on the questions, "would our fences work" and "would we see any Crawfish Frogs"? They did, and we did. In 2010, we asked ourselves questions such as "would we see as many frogs as we saw in 2009," "would we see the same frogs again,"

---

* *Rana sylvatica.*
† *Ambystoma texanum.*

"how much had they grown over the course of the year," "would females skip breeding," "would frogs be loyal to their 2009 breeding wetlands or would they move to another," and "would any of the juveniles we'd marked in 2009 show up to breed"?

We knew some of the frogs at our field site reasonably well from having filmed them at their primary burrows throughout the warm season in 2009. Several had behavioral quirks or burrows in odd locations, so we had naturally given them descriptive names. Once we named frogs, we found we began to care about them, a process we would best describe as the concern we showed for the species as a whole (which is good and forms the basis of conservation biology) filtered down to certain individuals (which can be bad if it introduces bias into your dataset—something we knew could happen and took care to avoid). Nevertheless, we were curious about the fate of these frogs we had come to know. So, in addition to the normal questions about what Crawfish Frogs in general were doing, prior to the 2010 breeding season, and for every breeding season after, we wondered whether these frogs we were familiar with had survived the winter, and if we would see them as they showed up to breed. When they did, it was cause for minor celebration, but our overwhelming emotions were admiration and awe that these frogs had made it through yet another winter (Figure 6.2).

FIGURE 6.2 A hardy Vanessa Kinney Terrell with an adult female Crawfish Frog. Two things to note about this picture—it's cold, and it's 2009. In 2009 we approached our study like ecologists—we handled animals with our bare hands. As soon as we discovered the presence of chytrid fungus on a subset of our animals, however, we began wearing nitrile gloves, a new pair for each animal. In pictures from 2010 on, we were always wearing nitrile gloves.

Vanessa Kinney Terrell's* drift-fence data clearly showed the pattern of Crawfish Frog attendance at breeding wetlands, which was repeated year after year (Figure 6.3). It went like this: males arrive, females arrive, females leave, males leave. There is a lot of slop in this sequence, but the progression makes sense. Sperm production is fast and energetically cheap. Males can make sperm while in their breeding wetlands, and we suspect that large, dominant males mate many times during a breeding season. In contrast, egg production is a slow, energetically expensive enterprise; females begin yolking up in the fall and because they lay all their eggs in a single breeding event are able to mate only once a year. The breeding strategy of Crawfish Frog males is best summarized as an "early and often" approach; for females, it is "choose well, dump, and run."

FIGURE 6.3 Vanessa Kinney Terrell finishing up her field notes after processing a Crawfish Frog captured in a pitfall trap entering one of our study wetlands. This photograph demonstrates our approach to this capture technique (described fully in Chapter 8). Note the association of bucket (pitfall trap) to drift-fence, the half-lid we secured to the outside lip of the bucket, the sponge we used as a float and for moisture during dry conditions, and the tomato stake we used to allow small mammals to escape.

---

* Shortly after Vanessa graduated and left our lab she married Zach Terrell. In her early publications she used her maiden name; in her subsequent publications she used her married name.

As a result, males tend to show up first and spend weeks in breeding wetlands, seeking mating opportunities, whereas females spend days, usually leaving during the first big nighttime rain after they've mated and deposited their eggs.

In ecology, everything has consequences, and there are repercussions to these different, male vs. female, approaches to breeding. For example, when male Crawfish Frogs begin their early spring breeding migrations, it is often cold and most snake species are still overwintering, but by the time females have stopped showing up and males are ready to leave, summer is looming, and snakes have become active, making post-breeding migrations predator dense, and therefore more hazardous. Females, on the other hand, often enter and exit breeding wetlands within a couple of days, at a time when snakes may or may not be active, depending on recent ambient temperatures. The risk of females being preyed upon by snakes is on average probably about the same as the risk to males—higher than the risk to pre-breeding males but lower than the risk to post-breeding males.

As an aside, this difference in pond residency times between males and females, noted above, is reflected in the data from our two approaches to sampling breeding adults. Our drift-fence data (which captured at least 99% of all breeding adults) showed a male:female sex ratio of 55:45; our funnel-trap data (with capture probabilities based on retention times) were strongly male-biased at 80:20 (see also Chapter 8). In wetlands we funnel-trapped, the number of females captured never came close to equaling the number of males captured, or the number of egg masses laid (= breeding females present), which meant we missed capturing a large percentage of breeding females.

The objective of pre-breeding migrations is to find wetlands that are, or will be (if you're the first frog in), inhabited by other adult Crawfish Frogs; post-breeding migrations simply get frogs home. Target size (large pre-breeding wetland target vs. tiny post-breeding burrow target) factors into the length of time these migrations take, so does noise. Big, calling-frog filled, noisy breeding wetlands are much easier to locate than small, quiet, primary burrows embedded in a tallgrass prairie glade. Jen's data showed that post-breeding migrations are longer than pre-breeding migrations because searching for their primary burrows takes Crawfish Frogs more time than searching for breeding wetlands. Time outside of burrows means exposure to predators, and exposure to predators increases the probability of being preyed upon.

Jen's discovery, on 1 April, 2010—that following breeding, Frog 160 returned to the same burrow she occupied the previous year—suggesting to us that Crawfish Frogs have a "home," was not an isolated incident. Following the 2010 breeding season Jen summarized the locations of her telemetered frogs, as follows:

- Northwest Nate's ♀ back in her [2009 primary] burrow
- 460 back in his [2009 primary] burrow

- 160 back in her [2009 primary] burrow
- 580 back in his [2009 primary] burrow
- Romeo in Juliet's [old] burrow
- Juliet [out] looking for [new] burrow
- 139 back in his [2009 primary] burrow
- 060 in his [2009 primary] burrow.

As Jen emphasized when we discussed this, the landscape in 2010 following the fall 2009 prescribed burn looked nothing like the landscape migrating Crawfish Frogs experienced in 2009. Since this burn occurred before the 2010 breeding season began, this meant either that Crawfish Frogs were using cues acquired during their 2010 pre-breeding migration to guide them back to their burrows during their 2010 post-breeding migration, or that Crawfish Frogs do not need vegetation characteristics (indeed, given how hot the burn was, that they didn't need vegetation at all) to guide them back to their home burrows.

We did not address the big question of which cues Crawfish Frogs use to guide their migrations, except to note that because Crawfish Frogs typically migrate at night during rains, they cannot use celestial cues—clouds block any view of the stars and planets. Jen found that about 80% of retreat sites used by Crawfish Frogs during their post-breeding migrations had been used by these same frogs during their pre-breeding migrations, a form of station-to-station, Pony Express-style, movement.

One quick story while we're on topic. In yet another example of exceptions providing insights, during the 2011 breeding season, the first Crawfish Frog to show up at Cattail Pond, on 28 February, was a female. That in itself was unusual. We knew from toe clip markings we had given her (explained below) that she had metamorphosed from Nate's Pond in 2009, which meant she was two years old and this was the first time she had bred. The question then became, since there were no male Crawfish Frogs calling in Cattail Pond, how did she know Cattail Pond was a Crawfish Frog breeding wetland? We could imagine three possible answers. 1) She got lucky—she was sexually mature, it was time to breed, she began migrating, then serendipitously found Cattail Pond. 2) She was guided by the sound of Spring Peepers, Chorus Frogs, and/or Southern Leopard Frogs, which were calling when she began migrating.* The challenge to this idea is that these other species call from more wetlands at our study site than Crawfish Frogs use to breed. If this idea was true, Crawfish Frogs would be present everywhere these other species call, but they are not. 3) She remembered and followed the direction of calling Crawfish Frogs from the previous year's breeding season. Specifically, we

---

* Wells (2007, p. 308) states, "Under some circumstances, heterospecific calls, or certain features of heterospecific calls, can elicit phonotaxis by females (Gerhardt 1994)."

suppose that following metamorphosis in 2009 she dispersed as a juvenile and ended up near Cattail Pond, overwintered in a crayfish burrow, and emerged early in the spring of 2010, still immature. She heard calling Crawfish Frogs at Cattail Pond, recognized them for what they were, and noted the direction. A year later she emerged from her burrow, now gravid and driven to breed. She remembered the direction of breeding calls from 2010 and vectored in the direction of Cattail Pond, becoming the first frog in.

From among these three scenarios, we feel the third is the most likely. Skeptics (there are always many and unless they are dogmatic we like them because they keep us honest) will ask how the brain of a frog, the size of a pea, can remember anything over the course of a year. It's a fair question. We counter by in-turn asking a question about a Crawfish Frog behavior we know to be true: how can a frog that has been in a breeding wetland for a month and preoccupied with mating leave that wetland and return to its burrow—smaller than a putting cup on a golf green—a kilometer away? For these post-breeding migrations to succeed, some form of memory must be involved. Males, who have spent up to a month in breeding wetlands, must have a memory that lasts at least this long. If a frog can remember the location (or, minimally, the direction and distance) of its burrow for a month, why can't it remember the direction of a breeding wetland for a year? Put this way, the question of Crawfish Frog memory, even if it is reflexive and not consciously considered, centers around how long a memory can last, not the animal's ability to remember.

We collected additional evidence of the cognitive awareness of Crawfish Frogs. As early as 17 October, 2009 we wrote in our field notes (with a prose as tortured as Le Conte's, Chapter 2): "Day activity different in burn, mostly night activity." Two years later, in late 2011, following a hot, mid-August prescribed burn that eliminated all vegetative cover from about 100 acres at our study site (the same burn where we counted and located crayfish burrows), we decided to determine whether these early casual observations were true or not. Nate used wildlife cameras to record the activity patterns and behaviors of Crawfish Frogs occupying the post-burned prairie and compared them to the activity patterns and behaviors of frogs occupying non-burned, normally vegetated prairie (Engbrecht and Lannoo 2012). We (mostly Nate) analyzed a total of 24,581 digital images representing six weeks of observations on eight Crawfish Frogs (four each in the vegetated and burned habitats). Nate found that while frogs occupying vegetated and burned habitats exhibited similar nighttime behaviors, in daylight Crawfish Frogs in vegetated habitats spent most of their time outside on their feeding platform, while frogs in burned areas spent most of their time at or in their burrow entrance. Further, frogs in burned areas emerged later in the day than frogs in vegetated areas. These behavioral shifts suggest that Crawfish Frogs in burned areas are aware they are exposed to predators in daylight, and alter their behaviors in ways that maximize concealment while still allowing them to feed.

One additional example from our field notes: on 7 June, 2010 Jen wrote:

> [there's a] Prairie Kingsnake [burrow] essentially between two Crawfish Frog [burrows]—101 and the telemetered guy. We don't see them (esp. 101) out [of their burrows] as much [as other Crawfish Frogs]. Wondering if presence of nearby snake makes Crawfish Frogs nocturnal.

Again, we do not know whether such behavioral shifts are the product of reflexes or some form of primitive cognition, but they demonstrate Crawfish Frogs have behavioral plasticity—that they do not employ a one-size-fits-all response to their ecological challenges. We'll examine further evidence of Crawfish Frog plasticity in the face of environmental variation in Chapter 9.

Harleys through Downtown Milwaukee 7

Crawfish Frog breeding calls are loud (over 107 dB at a distance of 1 meter (Gerhardt 1975)) and can carry a long way (Smith 1961; Williams et al. 2013) (Figure 7.1). One evening, from a hill about a kilometer away, listening to a wetland where, because we had just been there, we knew Spring Peepers, Chorus Frogs, Southern Leopard Frogs, and Crawfish Frogs were chorusing, we could hear only Crawfish Frogs. We have often wondered why Crawfish Frog calls are so loud, and the best answer we can give at this time is so that Crawfish Frogs occupying distant burrows can hear them.

Physicists tell us that the reason Crawfish Frog calls carry so much farther than the calls of other spring breeding frog species is that they have the lowest frequency. Low-frequency sounds travel as slow waves, which encounter much less air resistance than high-frequency fast waves, and therefore have staying power. In addition to being heard over great distances, this slow attenuation of low-frequency sounds means that calling frogs are difficult to locate. A single Crawfish Frog calling 10 meters (30 feet) away, and one calling 100 meters (300 feet) away, can sound about the same to the human ear. When we try to locate a calling male in a wetland (if you can locate choruses you know where to direct your search for egg masses), it is impossible for us to discern whether the frog is close by on the near shore, or distant on the far shore. It takes two of us, separated, to triangulate the frog's location. We were discussing this phenomenon during an evening social hour at a scientific gathering. One field biologist told us about the time he heard a Crawfish Frog chorus, parked his car and set out on foot to find it. He ended up crossing the next parallel road, a mile away (where he could have parked), before coming up on the wetland. When contemplating the secretive nature of Crawfish Frogs, add to the list that includes solitary burrow dwelling, the difficulty of locating calling males.

For all the energy Crawfish Frog males put into their breeding calls, they are often shy about making them. As a human approaches a typical twenty-first century Crawfish Frog chorus, males will abruptly stop calling and the

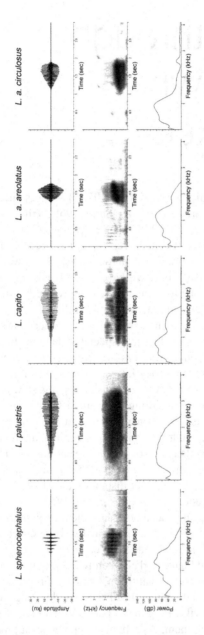

FIGURE 7.1 Comparative call characteristics of species in the subgenus *Nenirana*. Song oscillograms (amplitude), sonograms (frequency), and power outputs of Southern Leopard Frogs, Pickerel Frogs, Gopher Frogs, Southern Crawfish Frogs, and Northern Crawfish Frogs recorded at 15±0.5°C. This figure was first published in Lannoo et al. (2018) and is used with the permission of Leo Smith and the American Society of Ichthyologists and Herpetologists.

night will go silent. If the person then sits quietly near the wetland, in 10 or 20 minutes one male will begin calling, then another, and another, building until the full chorus resumes. Mike Redmer, now with the U.S. Fish and Wildlife Service, taught us a second way to experience Crawfish Frog choruses close up. Have two people approach, the chorus will go silent. Then have one person walk away. Crawfish Frogs cannot count, so the one person walking away eliminates in the frogs' cerebral cortices the perceived threat from the two people who approached. With the perception of threat removed, the full chorus soon resumes, and the person remaining is thrilled.

When only two Crawfish Frog males are calling, frogs tend to take turns, or alternate calls, a behavior termed antisynchronization. Antisynchonization is most often studied in frogs within a breeding wetland (Aihara et al. 2014), but on one lazy night in 2011, MJL heard a male at Nate's Pond alternate its calls with a male at Cabin Pond, half a kilometer away.

Over the eight breeding seasons of our study, we experienced on four occasions—once each in 2009 and 2011, and twice in 2016, all at Big Pond— Crawfish Frog choruses so large and intense that the frogs were oblivious to our presence. These big choruses sound like Harleys rolling through downtown Milwaukee. The two nights in 2016 were back-to-back. We know from our sampling that this chorus comprised over 100 breeding adults, and at any one time there could have been as many as 75 males calling. These Crawfish Frogs were so indifferent to our presence that we simply waded out into the wetland and grabbed them by hand. Historically, many Crawfish Frog choruses must have been this large and noisy,* but the diminishing size of many of today's populations make these large choruses increasingly rare.

Across a breeding season, Crawfish Frog chorusing levels vary with weather conditions, but at HFWA-W peak chorusing almost always occurred during a short, few-day-period between late March and early April, as follows (Williams et al. 2013). The beginning of breeding was typically marked by a gradual increase in chorusing levels as more males entered the wetland, interrupted by nights of little to no calling during cold, windy, or rainy conditions. Following peak breeding, calling activity for the season typically dropped off rapidly. During each night of calling, intensity increased during the first hour after sunset and, following a period of peak chorusing lasting about 1–2 hours, grew weaker as the night progressed, usually ceasing except for scattered calls until around 3:00 AM (Williams et al. 2013). There were exceptions. During a follow-up survey on 29 March, 2019, a warm, rainy night following an unusually cold spring, Nate and MJL experienced a Crawfish Frog chorus that lasted the entire night.

---

* For example, Cagle (1942) reports collecting 289 frogs from several small ponds in close vicinity.

In 2009, Jen and MJL were locating telemetered frogs and had been out all night. Around 5:00 AM, a Barred Owl called, which triggered howling in a pack of coyotes, which triggered calling in a single Crawfish Frog in Nate's Pond.

Similar to Gopher Frogs, where this behavior has been better documented, Crawfish Frogs will emit breeding calls underwater.* We have heard these calls a handful of times, and found that you have to be in or near the wetland in order to detect them. Roger Barbour was likely mistaken when he wrote, "They can call from underwater and in air, and an individual calling from under 18 inches of water can be heard at a distance of half a mile" (Barbour 1971, p. 112). We have asked experts about this underwater calling in both Crawfish Frogs and Gopher Frogs,† and there is general agreement that even under optimal circumstances (shallow water, favorable atmospheric conditions) submerged calls can be heard at distances of only a few hundred meters, if that.

In addition to producing breeding calls, Crawfish Frog males produce non-reproductive calls, which we call upland or burrow calls. Male Crawfish Frogs have long been known to call from their burrows (Smith et al. 1948, p. 609), but the function of these upland calls has remained a mystery. Calling is energetically expensive, and we were curious why Crawfish Frogs would call outside the breeding season. To address this question, we compared upland calls with breeding calls (Engbrecht et al. 2015). We found that upland calls differ from breeding calls in several important ways (Figure 7.2). First, upland calls are about a third as loud as breeding calls. Second, upland calls are lower pitched than breeding calls (690 vs. 810 Hz). Third, upland calls are drawn out (in technical terms have longer pulse periods and durations) compared with breeding calls. Interestingly, frogs can shift between upland and breeding calls abruptly, suggesting different functions for these vocalizations. Frog 26 (Romeo) shifted from producing upland calls to breeding calls within 2 hours on 11 March, 2013, as follows. Shortly after sunset, on 10 March, Romeo emerged from his burrow and a few hours later begin calling. He completed three calling bouts, beginning at around 1:00 AM and ending about 90 minutes later. The calls Romeo made during the first bout were pure upland calls. Calls he made during the second bout varied, alternating between upland and breeding calls. Calls he made during the third bout also varied, but in a clear direction; after initially demonstrating characteristics intermediate between upland and breeding calls, Romeo transitioned into producing pure breeding

---

* Dundee and Rossmam (1989, p. 108) "Francis Rose (pers. comm.) said that he had heard [Crawfish Frogs] singing from beneath the water."
† In an online listserve (sevosa@listserve.eku.edu) discussion that included J. Jensen, J. Palis, M. Baily, P. Frese, J. MacGregor, V. Meretsky, K. Buhlmann, M. Lodato, and M. Sisson.

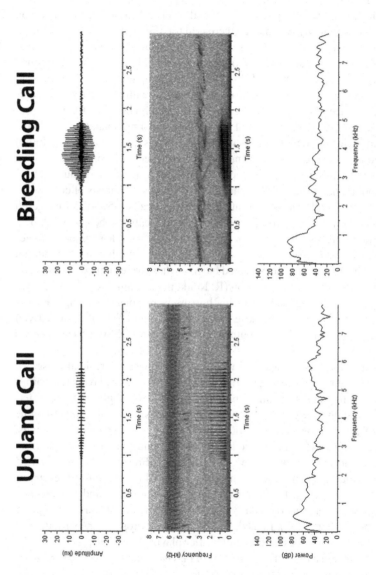

FIGURE 7.2 Upland vs. breeding calls in Crawfish Frogs. Waveforms, spectrograms, and spectrogram cross-sections of a typical Crawfish Frog upland call (left) and a typical breeding call (right). Ambient temperatures for both calls were 16°C. This figure was first published in Engbrecht et al. (2015) and is used with the permission of Leo Smith and the American Society of Ichthyologists and Herpetologists.

calls. Immediately after, he began his breeding migration, and the next morning we captured him entering Cattail Pond.

We found that following breeding and after returning to their burrows, Crawfish Frog males were quiet for several weeks, then, as the warm season progressed they slowly began, then increased their upland call activity. Calling rates rose gradually from no bouts in May to a rate of 0.01 bouts/hour in June; calling rates then abruptly rose to 0.06 or 0.07 bouts/hour in July and August. Rates peaked at a maximum of 0.1 bouts/hour in September. In October and November, Crawfish Frogs decreased their rates of calling as cooler fall temperatures reduced their overall activity levels (Engbrecht et al. 2015).

In our study, only 22 (40%) of the 55 post-breeding upland calling bouts produced by male Crawfish Frogs were spontaneous; the remaining 33 (60%) appeared to be triggered by a sound. Of the triggered calls, only five (15% of all triggered calling bouts) were associated with a natural stimulus such as rain, wind, thunder, the calls of other Crawfish Frogs, the howls of coyotes, or the honks of Canada Geese. Twenty-eight upland calls (85% of all triggered calling bouts) were immediately preceded, and appeared to be triggered, by a manmade stimulus such as an airplane, car, or truck (4-cycle engines), radiotelemetry receivers (the white-noise static background), or other forms of human disturbance. We located one occupied burrow when the resident Crawfish Frog called in response to a conversation (R. Ronk, pers. comm.). On 1 September, 2010, we filmed Romeo calling at his burrow in response to a single-engine airplane flying overhead, and have posted this video to YouTube, so that you might see it, too: https://www.youtube.com/watch?v=ojO5BrCEfsU&feature=youtu.be.

Because Crawfish Frogs appear to call in response to artificial noise, we were interested to see if we could elicit a calling response by playing recordings to males residing in their primary burrows. In 2012, we employed three different types of sounds originally recorded at our study site: a small, single-engine airplane flying overhead, a series of Crawfish Frog post-breeding upland calls, and a series of Crawfish Frog breeding calls. We conducted playback observations at the burrows of two male Crawfish Frogs (Frogs 26 [Romeo] and 65 [Private]—the only male frogs we had localized at this time) during seven different sessions on 4, 12, 17, 22, and 25 October, 1 November, and 3 December. Playbacks were performed on days when temperatures were warm enough for frogs to be active (>10°C). Indeed, we often saw the frogs on the surface next to their burrow entrances before and/or after playback sessions.

We presented a total of 218 minutes of playback sounds—36 and 182 minutes, respectively—at the burrows of Romeo and Private. At Romeo's burrow, we conducted 20 minutes of airplane flyover playbacks, 8 minutes of post-breeding upland call playbacks, and 8 minutes of breeding call playbacks. At Private's burrow, we conducted a total of 162 minutes of airplane flyover playbacks, 12 minutes of post-breeding upland call playbacks, and 8 minutes of breeding call playbacks. Private called in response to playbacks on three different occasions: once in response to the sound of post-breeding upland calling (12 October) and twice in response to the sound of an airplane (22 October, 3 December). Romeo did not call in response to any of the playback recordings and, therefore, for once in his long life, did not offer us any useful data. This exercise demonstrated that male Crawfish Frogs will call in response to conspecific upland calls as well as to artificial sounds, and confirmed the impression that we drew from analyzing 1,350 hours of audio recordings made from 2009–2011 at burrow entrances, as well as the video recording of Romeo calling in response to an airplane flyover in September of 2010.

We suspect that male Crawfish Frogs respond to airplanes, vehicles, radio-telemetry receivers, and human speech because the frequencies of these man-made calls overlap with the frequencies of Crawfish Frog-generated upland calls—that male Crawfish Frogs respond to these irrelevant (and irreverent) sounds because they perceive them to be the relevant calls of potential competitors.

Given all these observations, we return to our original question: why do Crawfish Frog males bother to call from their burrows? We considered several options, and the most likely appears to be territorial (burrow) defense. Calling to defend specific locations is a well-documented behavior in frogs. And, as we have seen above, because burrows assist Crawfish Frogs in deterring predators, having a well-fitted burrow likely means the difference between frogs like Romeo and Private living or dying—it is a resource worth defending.

The Circle of Life

8

At its core, wildlife management is nothing more than discovering where, when, and how death occurs, and attempting to prevent or reduce it. Therefore, to understand Crawfish Frog declines, it was Vanessa's job to collect the data that would allow us to pinpoint the sources of death, and to do this she had to understand how their populations are structured (demography). Her approach was a common one among amphibian biologists. She separated Crawfish Frog life history into five discrete stages—egg, tadpole, juvenile, first-year breeding adult, and older breeding adults—and examined mortality (or its cup-half-full inverse, survivorship) during each stage. Vanessa and her crew accomplished this by constructing the drift-fence arrays, mentioned above, around two major breeding wetlands, Nate's and Cattail Ponds (Big Pond was also a major breeding wetland but was too big and its margins too ill-defined to build and reliably maintain a fence).* These fences intercepted migrating adults first entering

---

* Constructing drift fences is hard work and expensive. To build our fences, which we erected at two wetlands—Nate's and Cattail—we used the following supplies: square 4-gallon buckets; tough, 4-feet high, erosion control fencing; 8-feet 2 × 2 boards; hardware cloth; swimming pool noodles; rectangular natural sponges; 6-feet tomato stakes; deck screws; staples; and 4-feet laths. For equipment, we used shovels, a portable drill with several extra lithium batteries, and a staple gun. To build the fence we began in the lab by first removing detergents and oils from the sponges. We soaked sponges for a week in water with soap remover, and rinsed four times, a day each time, in changes of fresh water. Before we headed into the field we cut the bucket lids, 2 × 2s, and laths in half, quartered the tomato stakes, and drilled three holes for screws along each lath. Around our study wetlands, approximately 5 meters from the water's edge, we dug a trench for the fence, and buried the bottom 6 inches of fencing in the trench. Every 10 meters, we dug two 1-feet by 1-feet holes, 18 inches deep and sunk the buckets, one bucket on either side of the fence. Every 6 feet or so along the fence, we pounded in a 2 × 2 for support, and we erected the fencing by pulling it up taut and stapling it to the boards. Then we secured the fence by screwing laths to the 2 × 2s. This is critical. We found that if we didn't add supporting laths, spring winds and thunderstorm gusts worked on the fencing fabric and started pulling out staples. When that happened big sections of fencing tore free from its stakes and flapped like a sail. Under these conditions, it didn't take much wind to destroy the integrity of the fence, and allow frogs to trespass. Once the fence was up and the buckets dug in and re-packed with soil, we finished the array by placing one sponge and one tomato stake in each bucket. The sponge provided moisture during dry conditions and a float when buckets flooded. The tomato stakes

then exiting breeding wetlands. As well, they intercepted newly metamorphosed juveniles as they exited wetlands and began dispersing into upland habitats. As Vanessa (and later, RMS) processed these frogs, they gave them unique markers—tags or clips—which allowed us to track their fate.

The concept of a drift-fence/pitfall trap array is simple—frogs migrating into a breeding wetland and encountering a *fence* will move (*drift*) along it, and if you dig *pits* at intervals they will *fall* into them, where they become *trapped* (Gibbons and Bennett 1974). A similar thing happens when frogs leave wetlands, moving in the opposite direction. Vanessa and her crew constructed our drift-fence/pitfall trap arrays in early 2009, and every year of our study we opened them to monitor breeding adults and metamorphosing juveniles (Figure 8.1). As a check on breeding activity, during the late winter, when it was still cold and in advance of any likely Crawfish Frog movement, we deployed Song Meter audio recording units. These recordings let us know whether males had entered breeding wetlands unexpectedly early, or if resident males occurred within the fence boundary, as occasionally happened. On 11 May, 2011, our field notes read "Tiny Crawfish Frog, 44 mm, 7 g, emerged from Nate's [Pond]. Must have overwintered or trespassed twice [once going out as a newly metamorphosed juvenile, a second time coming back in]—not likely when that small".

Animals that cross drift-fences without being detected are called trespassers. Trespassing regularly occurs with small, agile frogs such as Chorus Frogs, Spring Peepers, and Gray Treefrogs, but rarely with large, bulky frogs such as adult Crawfish Frogs or their tiny, 30 mm- (inch-and-a-quarter-) long juveniles. Using capture/re-capture data, Vanessa estimated a trespass rate for Crawfish Frog adults at our drift-fences of <1%. This low value meant that our sample scheme gave us, in essence, a population census (a count of every individual) rather than an estimate (a number from sampling, which ideally should be based on a randomized subset of the population).*

(in theory) allowed small mammals to escape. We then put half-lids on each bucket, open side nearest the fence, to protect trapped frogs from being preyed upon by raccoons, which learned to patrol our fence perimeters. In areas of fences prone to flooding, we used hardware cloth instead of fencing to allow water to flow through; water pressure against our erosion fence material collapsed the fence. We cut pool noodles lengthwise, on one side, and placed a noodle over the top of each hardware cloth section. As crazy at it seems, Crawfish Frogs can climb hardware cloth, but cannot then negotiate a pool noodle covering the top. We checked each bucket at least once a day, at dawn, because Crawfish Frogs migrate during the night. Also, because Crawfish Frogs migrate during and following warm rains, we checked buckets on rainy nights throughout the night. We would often drive up to our wetlands just as severe storms were clearing the area, and it was on these nights that we processed the most frogs (for example, 48 in one night in 2009, 43 in 2011, 86 in 2013, 87 in 2014, 84 in 2016, etc.).

* Mills (2007, p. 60) states: "Census is reserved for the special and unusual case where detection probability equals 1.0, that is, all animals were counted."

FIGURE 8.1   The drift-fence encircling Nate's Pond. Pitfall trap buckets were dug into each side of the fence at 10-meter intervals (one white bucket is visible along the fence in the lower right portion of the photo). At Nate's Pond, this fence was 280 meters long, so we employed 27 pairs of buckets. Note, along the fence to the left of the pitfall trap bucket is one of two areas where we replaced erosion fencing with hardware cloth to accommodate the flow out of Nate's Pond, which often occurred after heavy rains.

Most descriptions of drift-fence/pitfall-trap arrays fail to mention that by detaining animals, fences increase mortality among members of the population of interest by exposing them to predators. For Crawfish Frogs this happens in two ways. First, drift-fences delay frogs at the fences, where they are exposed to predators. Early in our study we found the carcasses of three Crawfish Frogs preyed upon while migrating into wetlands, as they were being detained outside our fences. (During frog breeding migrations, all a snake or a raccoon predator has to do is circle these fences to find frogs held up by the fence before they have a chance to move along the fence and fall into a bucket.) Second, drift-fences indirectly increase the exposure of frogs to predators by delaying post-breeding migrations from a time that is optimal for Crawfish Frogs to migrate (nighttime rains) to a time that is not (dawn or just after). Here's how this works.

Imagine a late evening downpour that triggers post-breeding migrations—frogs are in wetlands, they have bred, and now it's raining and time to go home. A little after dusk Crawfish Frogs exit wetlands for burrows that, at an average of 350 meters out, are a long night's trek away. Frogs leave the water and almost immediately encounter our drift-fence and fall in a pitfall trap. If we check traps eight hours later, at sunrise, process these frogs and release them outside the fence into a clump of Big Bluestem or Indian Grass, they must then wait until the next heavy nighttime rain to resume their migration. Prairie grasses do not provide the resistance to predators that wetlands or burrows afford, and the next nighttime rain might not occur for days or weeks. The longer Crawfish Frogs are out of their burrows the warmer spring becomes; with this warmth comes snake activity, and with this snake activity comes Crawfish Frog death.

Crawfish Frog idiosyncrasies do nothing to improve this scenario. Unlike Crawfish Frogs entering wetlands, or other amphibian species exiting wetlands, Crawfish Frogs leaving wetlands do not move along fences and fall into pitfall traps in the same way they do when entering wetlands. Instead, they try to work through the fence at the intercept between the fence and the straight-line vector to their burrow. Some frogs eventually encounter buckets and become trapped. The frogs that do not, persist until, as sunrise approaches, they give up, turn around, and head back into their wetland for another try during the next rainy night. Early on, we discovered that frogs that spent a night trying to work their way through a drift-fence could severely abrade their snouts (Heemeyer et al. 2010a). While it seems that this is a relatively minor injury, it must not be, because after we released these frogs, we never saw them again (i.e., they died).

To avoid this problem of drift-fence-caused frog retention leading to an increased probability of injury and death, we decided to monitor pitfall traps every rainy night, typically all night long, in addition to checking them at sunrise. We felt we owed these frogs at least this consideration.

In the four wetlands either too large (Big and New Ponds) or too small (Erosion Control and Nate's Jr. Ponds) to use drift-fence/pitfall-trap arrays, Vanessa instead used aquatic funnel-traps to sample Crawfish Frogs. The traps we preferred consisted of cloth mesh wrapped around a cylindrical wire frame, with concave funnels at both ends. The idea is that once frogs swim or crawl into these funnels, the funnels guide them into the trap where, because of the orientation and shape of the funnels, frogs usually do not escape. It is critical to offer these trapped animals access to atmospheric oxygen either by placing funnel-traps in shallow areas (with a portion of the trap above water), or by inserting into each trap a float, such as an empty 2-liter soda bottle, to ensure buoyancy. Without access to atmospheric oxygen, adult frogs, as well as trapped turtles and aquatic insects (including the magnificent Giant Water Bugs, Water Scorpions, and Dragonfly naiads), will drown (Klemish et al. 2013).

Vanessa, and later RMS, set 36 funnel-traps at Big Pond, 12 at Erosion Control, 12 at New Pond, and 12 at Nate's Jr., for a total of 72 traps. In each wetland, crews set traps provisionally, using calling locations or the previous years' breeding site(s) as an initial guide, then adjusting trap positions based on the location of egg mass clusters. As Bragg pointed out (Bragg 1953):

> In the production of eggs … these frogs are quite gregarious … wherever three or more clutches of eggs were found in a single pool or ditch, nearly invariably they were very close together, even in extensive pools where there were many other places compatible in all respects that I could discern to the egg-laying site.

This tendency for Crawfish Frogs to lay their eggs in clusters says something about their method of breeding (below), but more practically it indicated to us the specific location of breeding sites, which meant placing our traps at these areas caught breeding adults. In 2014, RMS set traps in Big Pond based on the previous year's breeding location, but after two days caught no animals, even though she had heard calling from the general area. After checking where egg masses had been deposited she moved our traps ~8 meters (25 feet) to the east and immediately began capturing Crawfish Frogs. Over the next few weeks her crew caught 55 breeding adults.

Once Vanessa, RMS, or one of their crew captured an adult Crawfish Frog in a bucket or funnel-trap, they put on nitrile gloves, removed the frog, and sexed it. They then swabbed the animal for diseases, scanned it for a pit tag ("pit" is an abbreviation for "passive integrated transponder," essentially a glass-encased bar code inserted subcutaneously that allowed us to individually identify each animal) (Figure 8.2), and checked for juvenile toe clips (our system of two toe snips—one indicating year, the other wetland—that we subsequently used to identify where and when we had seen this animal as a post-metamorphic juvenile; we saved these toes and later used them for genetic analysis (Nunziata et al. 2013)). Then, Vanessa or her crew measured the length (in millimeters) and weight (in grams) of the frog, and inserted a pit tag if one was not present. If the frog was captured at a fence they released it on the opposite side, either in the water when it was heading into a wetland during its pre-breeding migration, or under a dense clump of prairie grass when the frog was beginning its post-breeding migration. If the frog was caught in a funnel-trap, they released it in a nearby cattail stand out of harm's way, where nobody would step on it as they checked the remaining traps.

The point to all this breeding is to produce Crawfish Frog eggs, so we surveyed each breeding pond every morning for newly laid egg masses. Vanessa, and later RMS, placed a flag with the date at each mass, and then monitored them until their eggs hatched.

In order to present a sense of what one of our field seasons was like, we include here a sample from our 2014 field notes. By 2014, during Crawfish Frog breeding activity, we processed frogs using two teams. One would check

FIGURE 8.2   The belly of a Crawfish Frog male showing the 10 mm, subcutaneous pit tag we had previously inserted, which at this point had settled into the ventral groin region.

Nate's, Erosion Control, and New Ponds; the other would check Cattail, Big, and Nate's Jr. Each team had a separate field notebook, so no single notebook contained all of our field data. Each day after we came in from the field, RMS collected the notebooks, scanned the data pages, and entered the data into what became a huge spreadsheet that recorded every frog we ever encountered. Also, each day, Rochelle backed up these data and took the hard drive home with her. Here is a sketch from MJL's field notebook that will offer a sense of what a Crawfish Frog breeding season was like.

14 March, 2014: First burrow calling.

17 March, 2014: Put up drift-fences at Nate's and Cattail. Song meters at both [wetlands] recorded a couple of [calling] Crawfish Frogs on both Friday and Saturday. Calling light. Total amount of calling ~15 minutes both wetlands both days.

20 March, 2014: Cold—some ice in buckets but not as much as two nights ago.

21 March, 2014 AM: Cattail: Light ice. One [large] frog in Bucket 2 going in. Trapped last year at Big then at Cattail. 106 mm, 139 g.

Found three calling males at Erosion Control—*west* side.

22 March, 2014: Last two days:

15 males
11 new animals
4 recaps
All recaps [were] from Big

Strategy of going out last night and seeing where frogs at Erosion Control and Big were calling and shifting traps [to area of calling] was solid.

23 March, 2014: To date:

17 Crawfish Frogs, all males
13 new, 4 recaps
No egg masses

24/25/26 March, 2014: Cold, ice as much as 1 cm [on wetlands]. Rochelle could walk on ice.

28 March, 2014:

29 [animals]—10 new, 19 recaps
Total: 49 [animals]—24 new, 25 recaps

29 March, 2014: First day we got a bunch of *Cambarus* [upland crayfish] [in buckets].

30 March, 2014: Ice—clear, cold.

Total [Crawfish Frogs] to date:
78 frogs: 68 M, 10 F
42 new: 39 M, 3 F
36 recaps: 29 M, 7 F

1 April, 2014: Huge day: 62 Crawfish Frogs in: 40 M, 22 F, 30 new to us.

Grand total:
127: 96 M, 31 F
72 new: 56 M, 16 F
55 recaps: 40 M, 15 F

Egg masses discovered today:
Big: 8 more
Nate's: 2
Cattail: 5
Nate's Jr.: 3

2 April, 2014: Out from 3:00–4:00 AM. Rain. Trouble with my pit tag reader (dim).

Big day: 58 total Crawfish Frogs: 39 M, 19 F.
18 frogs new to us.
25 recaps this year.
15 recaps previous year.

Grand total (so far):
160 total: 114 M, 46 F
90 new: 66 M, 24 F
70 recap previous years: 48 M, 22 F

3 April, 2014: Violent thunderstorms overnight. RMS out at 3:00 AM. RMS says at least 75 frogs.

In fact, 84 frogs: 60 males, 24 females
34 new to us.
25 recaps this year.
25 recaps previous years.

Grand total (so far):
219: 157 M, 62 F
124 new: 92 M, 32 F
95 recap: 65 M, 30 F

4 April, 2014:

Today, 47 animals: 29 M, 16 F, 2?

19 new to us: 11 M, 6 F, 2?
24 recaps this year (17 M, 7 F)
4 recaps previous years: 1 M, 3 F

Grand total (so far):

243: 169 M, 72 F, 2?
144 new: 103 M, 39 F, 2?
99 recap previous years: 65 M, 34 F
70 egg masses identified

5 April, 2014 AM: Romeo left Cattail.

Today, 29 animals: 20 M, 9 F
14 recaps this year: 11 M, 3 F
6 recap previous years: 3 M, 3 F

Grand total (so far):

258: 177 M, 79 F, 2?
153 new: 109 M, 42 F, 2?
105 recaps previous years: 68 M, 37 F

Egg mass counts:

Big: 25
Cattail: 26
Nate's: 21
Erosion: 10
Nate's Jr.: 8
New: 4
_____
Total: 94

Juvenile recaps new to us [so far, this year]:

2010 Nate's: 12
2011 Nate's: 28

6 April, 2014: Cold again, frost, ice

11 Crawfish Frogs: 8 M, 3 F
2 new to us: 2 F
8 recaps this year: 8 F
1 recap previous years: 1 F

Grand total (so far):

261: 178 M, 81 F, 2?

155 new: 109 M, 44 F, 2?
106 recaps previous years: 69 M, 37 F

7 April, 2014: Brought Romeo in today. Found him lethargic at the fence near Bucket 6. Put him in the incubator and he couldn't take the heat. Was in for a couple of hours only [and appeared dead]. Did revive him—heartbeat and shallow breathing, but he never came all the way back. Put him in MS 222 [anesthetic] for half an hour, then 10% formalin [to fix and preserve him]. Texted Jen to let her know.

New pit tags came in today, which was big, we were down to about 13.

8 April, 2014: Rain. Out at 2:00 AM. Many frogs starting to head out. Worked until 3:15 AM, waited out a storm until 4:15 AM, then worked until dawn.

11 April, 2014: Out at 1:45 AM all night. ~90 frogs, mostly exiting.

12 April, 2014: Chytrid frogs starting to come in: 2 M, 1 F.

13 April, 2014: Female chytrid frog died.

14 April, 2014: Out overnight again.

Summary of 2014 [Crawfish Frog] Breeding Season:

Huge # of animals
Male-biased
Kicked our butts
Romeo died
Immensely rewarding [to see so many frogs]
Exhausting
Management [recommendations] working
A little tired, burned out, satisfied

In addition, we include here two pages from RMS's field notebooks, with an interpretation of what our shorthand meant (Figure 8.3). The left-side page records frogs captured entering Nate's Pond on 3 April, 2014. The right-side page records frogs captured exiting Nate's Pond during the night of 11 April, 2014.

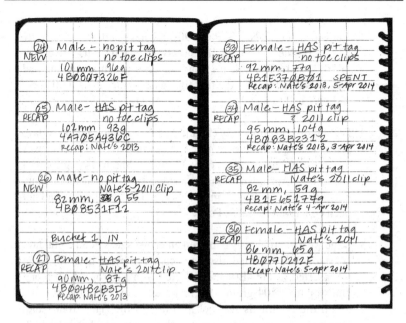

FIGURE 8.3 Two pages from RMS's field notebook. The left-side page records frogs captured entering Nate's Pond on 3 April, 2014. The right-side page records frogs captured exiting Nate's Pond during the night of 11 April, 2014. We explain our shorthand for each entry in the text.

24 Male captured entering Nate's Pond. He was a big frog, 101 mm long, weighing 96 grams, with no pit tag and toe clips, which meant he had not been conceived at Nate's Pond, and we had not seen him before. We gave him pit tag 4B0807326F and released him in Nate's Pond.

25 Male we had previously captured at Nate's Pond in 2013, when we had given him pit tag 4A705A436C. He was a big frog, 102 mm long, weighing 93 grams. He had no toe clips, which meant he had not been conceived at Nate's Pond.

26 Male captured entering Nate's Pond. He was a small frog, 82 mm long, weighing 55 grams. He had no pit tag, which meant we had not seen him before as an adult. He had Nate's Pond 2011 toe clips, which meant he had been conceived at Nate's Pond and was breeding there for the first time as a 3-year-old. We gave him pit tag 4B08531F12 and released him in Nate's Pond.

27 Female entering Nate's Pond 90 mm long and weighing only 87 grams. She had pit tag 4B08482B3D, which we had given her in 2013, and a 2011 Nate's Pond toe clip, which meant she had bred at Nate's Pond for the first time as a 2-year-old, and this year had returned to breed.

33 Female we had previously captured entering Nate's Pond six days prior (on 5 April) when we gave her pit tag 4B1E370B01. She was 92 mm long and weighed only 77 grams. She had weighed 104 grams on 5 April, which meant she had bred in Nate's Pond and her egg mass weighed 27 grams. She had no toe clips, so had not been conceived at Nate's Pond.

34 Male we had previously captured entering Nate's Pond eight days prior (on 3 April). Had also been captured by us in 2013, when we inserted pit tag 4B083B2312. He was a big frog, 95 mm long and weighing 104 grams. He had a questionable 2011 toe clip. (With these animals we took pictures of the right front hand, where a toe clip would indicate cohort year, and left hind foot, where a toe clip would indicate cohort pond, and made an assessment later in the warm and dry comfort of our lab.)

35 Male we had previously captured entering Nate's Pond on 4 April, when we inserted pit tag 4B1E651749. He is a small frog, 82 mm long and weighing 59 grams. He had a Nate's Pond 2011 toe clip, which meant he was three years old and been conceived at Nate's Pond.

36 Female we had previously captured entering Nate's Pond six days prior (on 5 April) when we gave her pit tag 4B077D292F. She was 86 mm long and weighed only 65 grams. She was spent, and had bred in Nate's Pond. She had 2011 Nate's Pond toe clips, so she was 3 years old and been conceived at Nate's Pond.

About a month and a half after breeding ceased, we began monitoring the development of Crawfish Frog tadpoles using funnel-traps, which we set out overnight in breeding wetlands. Feeding tadpoles swam into these traps, and the next morning one of us examined them for overall health and developmental stage. Once tadpole hind limbs became fully developed and front limbs were visible under the skin behind their heads, we began watching weather reports to know when to anticipate newly metamorphosed juveniles. Every morning at dawn, we checked fences for newly metamorphosed Crawfish Frogs. (If we checked fences at night, we risked stepping on and crushing these tiny, camouflaged animals.) As mentioned above, we toe clipped all dispersing juveniles using a right-hand clip to indicate year and a left-foot clip to indicate pond. We swabbed the first juvenile removed from each bucket to assess disease, and measured and weighed every juvenile. At the peak of metamorphosis, when hundreds of juveniles were leaving wetlands every night, even with all hands on deck, we were forced to scale back the amount of data collected in the name of animal health; these delicate little froglets will die if left long in pitfall traps exposed to the heat of a southern Indiana July afternoon. During these times we toe clipped each juvenile, but we only collected length and weight data on every tenth animal.

Once tadpoles began metamorphosing, the first big rain produced the largest pulse of juveniles, numbers then tapered off with subsequent rains. When we no longer captured Crawfish Frog juveniles emerging after heavy rains, and our funnel-traps no longer collected tadpoles in these wetlands, we snapped full lids onto our pitfall trap buckets and rolled down and secured our fences for the year.

Here's a quick summary of our data on breeding adults over the course of our study (2009–2016). Total numbers: 1,102 breeding adults consisting of 729 males, 371 females, and 2 unsexed frogs. At Nate's and Cattail Ponds combined (where we assume a <1% trespass rate), we captured a total of 338 males and 273 females, a slightly male-biased 55:45 sex ratio. Breeding adults averaged 92 mm and 97 mm (males and females, respectively), 93 g and 82 g (males and females), and average egg mass weight was 29 g (weight of post-breeding females subtracted from the weight of pre-breeding females). The longest frog was a gravid female at 120 mm (183 g, unknown age). The heaviest frog that we captured was a gravid female at 193 g (116 mm, unknown age). The smallest breeding male frog was 61 mm and 33 g (two-years-old); the smallest breeding female was 71 mm, 42 g (gravid), and 31 g (spent, unknown age).

While we could track individual adult Crawfish Frogs using pit tags, our assessments of juveniles had to necessarily be done by cohorts—by pond and year class—as indicated by toe clips. Below, we present our juvenile data for Nate's Pond to give some idea of the variation in numbers, size, and length of larval stage (time to metamorphosis). No frogs metamorphosed from Nate's in 2012 (due to drought) or 2015 (due to predation).

COHORT YEAR	# JUVENILES PRODUCED	AVG SVL (MM)	AVG MASS (G)	DAYS TO METAMORPHOSIS
2009	286	34±0.2	4.6±0.07	90±0.5
2010	2,103	33±0.1	3.4±0.03	82±0.3
2011	3,122	31±0.1	2.5±0.02	97±0.5
2013	8	33±1.7	2.9±0.70	116±8.6
2014	844	30±0.2	2.5±0.04	71±0.3

Body Language   9

Human-mediated climate change offers perhaps the greatest challenge to life on Earth,* and assessing the impacts of a shifting climate on species, especially threatened and endangered species, has become a top priority for conservation biologists. By affecting seasonal temperatures and precipitation (Environmental Protection Agency 2016), climate instability has the potential to dramatically alter the ability of animals to carry out critical aspects of their life history or natural history—factors necessary for individual health and population persistence (Ordonez et al. 2016). Amphibians, with physiological limits and behavioral responses tied to predictable temperature and precipitation expectations, appear to be especially vulnerable to climate instability (Corn 2003).

Spring breeding amphibians, in particular, may be severely threatened by environmental destabilization (Walls et al. 2013). These frogs and salamanders rely on temperature and rain cues to stimulate breeding activities. Following breeding, their aquatic larvae continue to depend on reliable periodic rainfalls to maintain water levels in seasonal/semi-permanent wetlands. Either precipitation extreme can result in catastrophic recruitment failure: too little rain early eliminates breeding triggers; too little rain late and wetlands dry before larvae develop sufficiently to metamorphose; too much rain, especially in torrents, and surface water transports predatory fishes into these otherwise fish-free basins. String enough of these immoderate years together and populations can collapse.

Such catastrophic responses to extreme weather events are common and have been well documented. But we can also ask whether there are less dramatic short-term effects of climate variation that impact populations in the long term. Such effects must be considered on a species-by-species basis. For example, in southwestern Indiana, Southern Leopard Frogs and Crawfish Frogs historically bred simultaneously in the early spring. However, in response to persistent warmth, southwestern Indiana Southern Leopard Frogs have begun breeding in the fall (Stiles and Lannoo 2015), while Crawfish

---

* https://www.ipbes.net/news/ipbes-global-assessment-summary-policymakers-pdf.

Frogs have not—an example of climate change driven community disassembly (Chan et al. 2016; Ordonez et al. 2016).

We addressed the response of Crawfish Frogs to short-term variations in temperature and precipitation driven, in part, by climate change, by addressing five questions: 1) Is there evidence for the timing of breeding shifts in response to drought (i.e., warmer, drier) conditions? 2) Do drought conditions affect adult survivorship? 3) Do drought conditions affect body mass? 4) Do drought conditions affect the number of eggs females produce? And finally, 5) because fecundity is indeed affected, what is the extent of this impact, measured in terms of estimated adult recruitment to the population?

1) Did we find evidence that the timing of breeding shifted in response to drought (i.e., warmer, drier) conditions? No. Crawfish Frogs did not use temperature as the primary cue triggering breeding. Once the frost is out of the ground, Crawfish Frogs emerge from their burrows and become active at their burrow entrances, but they migrate only during or after heavy nighttime rains, which can occur a week or more after frogs have become active. Droughts, by definition, mean fewer rains, so dry conditions may in fact delay pre-breeding migrations. After Crawfish Frogs are in breeding wetlands, longer intervals between rains may keep them there long past breeding. This creates a problem. Post-breeding migrations occur during warmer temperatures, when snakes have come out of hibernation and are actively feeding. Snakes are the chief predators of Crawfish Frogs, and the longer into the spring frogs remain in breeding wetlands, the more vulnerable they are to snake predation.

2) Did drought conditions affect adult survivorship? No, in fact, the opposite appeared to occur; our data suggested flooding following too much rain affected Crawfish Frog survivorship (Figure 9.1). It may seem counterintuitive that Crawfish Frog survivorship was reduced during the wettest years—most amphibians are, directly or indirectly, dependent on water. But the behavior of other amphibian species cannot always be generalized to Crawfish Frogs. Frequent rains during wet years flood Crawfish Frog burrows, which can be fatal. In particular, during the winter, heavy rains frequently precede cold fronts, and hard ice formed on recently flooded burrows will trap and drown Crawfish Frogs (Heemeyer and Lannoo 2011). Further, Crawfish Frogs will temporarily abandon submerged burrows, increasing their risk of predation. Crawfish Frog defensive behaviors and perhaps their head shape are tied to burrow occupancy, and Jen has estimated that a Crawfish Frog away from its burrow is 12 times more likely to be preyed upon than a frog at its burrow (Heemeyer and Lannoo 2012).

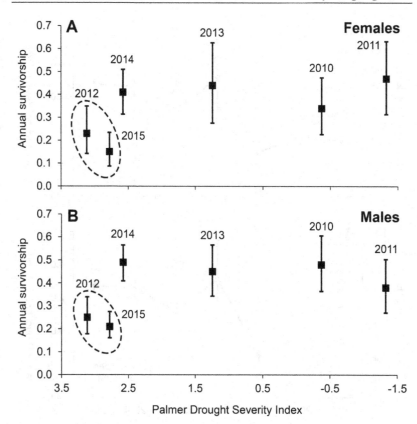

FIGURE 9.1 Drought (negative Palmer Drought Severity Index values) was not correlated with annual survival in either Crawfish Frog females or males at our study site from 2009–2015. However, wet years appear to have had a minor negative effect on survivorship (see circled values). This image was originally published in Lannoo and Stiles (2017) and is used with the permission of Leo Smith and the American Society of Ichthyologists and Herpetologists.

3) Did drought conditions affect body mass? Indeed, they did. Body mass was negatively affected by drought in both males and females (Figure 9.2). Presumably, this was due to the reduced availability of invertebrate prey and increased physiological stress due to elevated temperatures and limited potential for water intake. Crawfish Frogs are active any time of the year under favorable conditions—they do not exhibit any form of hibernation or winter torpor—and activity uncoupled from prey availability can reduce body mass values. Reduced body mass values may also be due to a frog's increased activity due to increased temperatures.

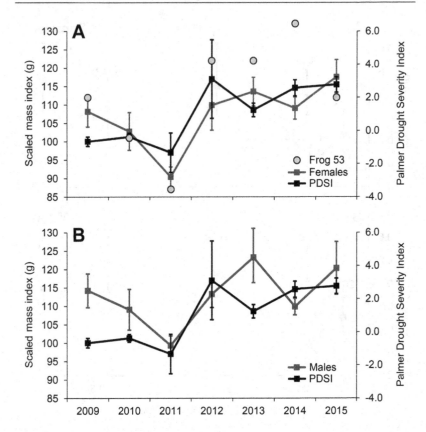

FIGURE 9.2 Drought prior to breeding influenced Crawfish Frog body condition (scaled mass index) of both (A) gravid females (n = 356) and (B) males (n = 730). We calculated SMI values for each frog entering breeding wetlands at our study site from 2009–2015. Frog 53 (Corner Burn Chick) was the only adult female captured every year of this study; her annual SMI values are plotted as circles (A) for comparison with the means of all captured frogs. This image was originally published in Lannoo and Stiles (2017) and is used with the permission of Leo Smith and the American Society of Ichthyologists and Herpetologists.

4) Did drought conditions affect the number of eggs females lay? Yes. We estimated that the difference in a female's clutch weight between the wettest and driest years of our study translated into a reduction in the average number of eggs laid per female of at least 2,500 (roughly 30%).

5) And finally, because fecundity was affected, we asked what is the extent of this impact, measured in terms of estimated adult

recruitment to the population? The numbers were staggering. Using calculated survivorship estimates by life history stage for Crawfish Frogs (detailed in Chapter 11: 98% survivorship to hatching; 1% larval survivorship; and 3% juvenile survivorship), in a population of 350 Crawfish Frogs, a loss of 2,500 eggs per female translates into an estimated loss of over 130 adult frogs recruited into the population.

These results suggest amphibians are much more sensitive to normal fluctuations in temperature and moisture availability than past workers have appreciated. Given this, we have suggested that collectively, we don't yet know enough amphibian biology to make accurate predictions about the future of amphibians in the face of increased climate variability (Lannoo and Stiles, in press).

When MJL assembled the *Amphibian Declines* book (Lannoo 2005), which contained over 700 manuscript pages of citations, he was struck by how few species represented the bulk of the literature. At the time he remembers thinking about the 80:20 rule—that 20% of the species had generated 80% of the literature.* Because this observation was made on North American species, where amphibians have likely been studied more than any other place on Earth, it's likely that over 95%, perhaps 99%, of amphibian species globally are deficient in the sort of data that are likely to inform the outcomes of conservation challenges.

Amphibian complexity in response to climate change has been noted before (for example, Muths et al. (2017)) as has physiological plasticity in the face of increasingly variable weather patterns (Gunderson and Stillman 2015; Gunderson et al. 2017). But, we offer here that the extent of this variability, as evidenced by the extreme sensitivity of Crawfish Frog BMI to temperature and moisture conditions, has not been fully appreciated, because nobody has collected the type of data that addresses this question. Nobody collected these data because in the golden age of natural history there weren't enough naturalists with enough instruments to collect them. And while today's naturalists have these instruments to collect these data, instead they tend to be inside sitting at their desks constructing models. (In fact, our models have become so sophisticated, we can now predict when they will fail (Woodin et al. 2013).)

There are, intermittently, calls to get back into the field. Over 20 years ago, Reed Noss published a paper entitled "the naturalists are dying off" (Noss 1996), and there have been more recent calls to get back into the field (Ferreira and Ríos-Saldaña 2016), but the truth is research careers are built on journal impact factors, and the highest impact journals favor modeling over collecting natural history data. As Harry Greene has pointed out, "If young naturalists become scientists, however—and this is truer now than ever before—acclaim

---

* Meta-studies such as Earl and Whiteman (2015) demonstrate the same trend.

likely flows from generalizing [i.e., models] than gathering [natural history] facts" (Greene 2013). But, as the legendary writer Jim Harrison answered, when asked why he had never accepted any of the cushy academic jobs he had been offered, "Somebody's got to stay outside" (Bissell 2001).

The world of science has become an odd place. If we accept the words of George E. P. Box, who should know: "all models are lies; some of them are useful" (Box and Draper 1987). And if we accept the words of Fortey that facts are truths that endure (Fortey 2008), then we are forced to conclude modern science places more stock in the world of transient lies than in the world of enduring truths. We suppose in this regard science is following politics, and probably society in general. Thomas Henry Huxley's quip about nasty, ugly little facts destroying beautiful hypotheses is no longer true. Today, ideas and ideologies may be more immune to contrarian facts than perhaps at any time since the Dark Ages. The idea of facts trumping models is not exactly dead, though. Unusual cold weather events such as polar vortices have driven a shift in terminology from "global warming" to "climate change."

Despite all this, models are the modern way of the world, and if we want our models about the fate of amphibians to be accurate they must incorporate as much data as possible on the physiological, behavioral, and ecological responses of every species to climate variability. The only way we get these data is to go outside and collect them. Until we do this, our models will be less than useful lies. With this in mind, we offer some perspective by Richard Nelson, who spent a year living with the Native Athabaskan Koyukon people in Alaska, and wrote about it in his book *The Island Within* (Nelson 1992, pp. 45, 161).

> The more people experience the repetitions of events in nature the more they see in them and the more they know, but the more they realize the limitations of their understanding ... I believe that Koyukon peoples' extraordinary relationship to their natural community has emerged through this careful watching of the same events in the same place, endlessly repeated over lifetimes and generations and millennia. There may be more to learn by climbing the same mountain a hundred times than by climbing a hundred different mountains ... Among the Koyukon people, I experienced an attitude quite different from that which prevailed in academic science. Elders ... carried their vast and insightful knowledge of the natural world with great humility. I never heard them speak of how much they knew, but how little, and of how much there was to learn, how difficult it was to understand even the smallest mysteries around them.

Feast or Famine 10

The reason the body mass indices of Crawfish Frogs change across years (Chapter 9) is that their bodies respond to environmental conditions within years. Our study site approaches the northern extreme of this species' range, which means that for half-a-year, from mid-October (the average time of first frost) until mid-April (post-breeding), Crawfish Frogs do not eat, or eat only lightly. During the fall and winter months, they are sporadically active and insect prey items are sparse. Prior to breeding, if temperatures are warm and the frost melts before a hard nighttime rain, Crawfish Frogs will be active outside their burrows and feed prior to breeding, but again pickings appear to be slim. During the breeding season, frogs are preoccupied with reproduction, and prey remains sparse. Available prey items early in the season are restricted to small insects and earthworms, and we're not sure if Crawfish Frogs will feed on earthworms (none of our mid-summer video clips of feeding bouts showed them taking earthworms). Our wildlife cameras and video recordings reveal that once Crawfish Frog adults return to their primary burrows following spring breeding they are active and feeding every time environmental conditions allow.

Prey availability changes as the warm season progresses, with larger insect prey becoming available (for example, dragonflies and damselflies, grasshoppers, crickets, locusts, and cicadas). Further, prey availability changes between day and night (for example, grasshoppers are active during the day while crickets are active at night). In addition to insects, Crawfish Frogs likely also feed on smaller frogs such as Chorus Frogs, Spring Peepers, Cricket Frogs, and the newly metamorphosed juveniles of Southern Leopard Frogs and Green Frogs; they could also be cannibalistic and feed on juvenile Crawfish Frogs.

As prey become available, Crawfish Frogs gorge themselves. This gluttony coupled with the relative immobility of Crawfish Frogs (at their primary burrows Crawfish Frogs move greater distances vertically within their burrows than they move horizontally across the landscape) means they gain weight fast—as you would expect from a couch potato. Examining our digital camera images, we can always tell it's late summer when Crawfish Frogs look as if they've swallowed a golf ball. We call these frogs "Buddha bellies," because they resemble the obese images of the spiritual leader.

Crawfish Frogs must binge during the six months when they can feed, because they face an upcoming six months of fasting. There are four important points about this non-feeding period to know, as follows.

1) During this time, Crawfish Frogs do not save energy by sleeping or hibernating—they are awake and at least semi-alert in the chamber at the base of their primary burrows—so they burn calories faster than they would if they were shut down.
2) During this time, females are forming eggs—5,000 to 7,000 of them, which is energetically expensive.
3) At the end of this period, Crawfish Frogs undertake their long breeding migrations (given that at our study site the mean distance of burrows from breeding wetlands is 350 m, the total average forth-and-back distance of breeding migrations is 700 m).
4) At the end of this period, and following their pre-breeding migration, males must produce loud (107 dB), energetically demanding breeding calls for weeks on end.

In short, Crawfish Frogs must gorge themselves when they can, because they never go completely quiescent, and their most energetically demanding time of year—breeding and its associated behaviors—occurs at the tail end of their six-month fast.

It would be interesting to compare annual activity cycles of Southern Crawfish Frogs, which do not experience true wintertime conditions, with the Northern Crawfish Frogs we studied. Southern Crawfish Frogs should be able to feed most of the year, which would suggest they should have larger body sizes. But they don't. As we mention elsewhere, breeding Southern Crawfish Frogs are small enough that they would be considered juveniles if encountered at our Indiana study site. This observation suggests metabolic demands are higher in warmer climates, and that life in this biochemical fast lane hampers attempts by Crawfish Frog adults to stockpile energy reserves.

Who Lives, Who Dies, and Why? 11

It wasn't until the twenty-first century, beginning with Mike Redmer's graduate work at Southern Illinois University (Redmer 1999, 2000), that questions concerning Crawfish Frog population biology began to be answered. Using a histologically based technique called skeletochronology, where a researcher counts annular bone rings as one might count tree rings, Redmer found that male Crawfish Frogs first bred at 2 years old, females at 3. Redmer also found that the majority of breeding Crawfish Frogs were either 3 or 4 years old, and that none of the 59 breeding adults he examined was more than 5 years old. (Our data, from a much larger sample size, indicate some Crawfish Frogs can live at least ten years.) Redmer also examined the association between female size (snout-vent length [SVL]) and fecundity (egg number) and found a correlation—larger females lay more eggs, and the relationship is linear. Vanessa first quantified, then exploited this connection (because Crawfish Frogs are state endangered in Indiana and could not be sacrificed, Vanessa could not kill and dissect animals, and therefore did not repeat Redmer's Illinois study). Using Redmer's data, Vanessa calculated the following regression equation:

$$\text{Clutch size} = -10{,}974.3 + 172.4 * \text{SVL}$$

To use this equation, all you have to do is plug a female's body length into the right-side variable (SVL) and do the math to get an estimate of the number of eggs this frog had laid. Between her drift-fence data and this equation, Vanessa had the tools she needed to work out the population biology of Crawfish Frogs. This is how she did it.

Every breeding season at Nate's and Cattail Ponds, Vanessa (and later, RMS) generated a census of the number of females who had bred there, and how big they were. She then calculated an average body length for females breeding at each pond. These data were facts. Vanessa then plugged the average body length of these females into the Redmer-derived equation to calculate the average number of eggs laid by each female (clutch size). She then multiplied this average clutch size by the number of breeding females to compute the total number of Crawfish Frog eggs laid in each pond for that breeding season. Using this technique, Vanessa calculated the number of eggs laid each year in each of our drift-fenced wetlands. These data were estimates.

After Crawfish Frog eggs hatched, which took about a week or two depending on wetland water temperatures (warm temperatures accelerate development and reduce the time to hatching), Vanessa and her crew gathered egg mass jellies and counted the number of un-hatched eggs, which she averaged across the number of egg mass remnants they'd sampled. She then divided this number by the average number of eggs/clutch to get an assessment of embryonic mortality, which she estimated to be about 2%. Finally, Vanessa multiplied the inverse of this number, 98% (0.98) by the estimated number of eggs laid to get an estimate of the total number of Crawfish Frog hatchlings (initial number of larvae) in that pond for that year.

By this point, we may have lost the non-mathematical among you, so here's an example.

In Nate's Pond, in 2009, the average SVL of breeding females was 100.3 mm. Using the Redmer-derived regression equation, Vanessa calculated the average clutch size for each female to be 6,320 eggs. Multiplying this estimate by the number of breeding females (32), she calculated that the total number of eggs laid was 202,226. Finally, multiplying this number by her estimate of embryonic survivorship (98%), Vanessa estimated that the total number of Crawfish Frog hatchlings at Nate's Pond in 2009 was 198,181.

Three months later, at her drift-fence, Vanessa and her crew captured 284 newly metamorphosed, dispersing Crawfish Frog juveniles leaving Nate's Pond. Dividing this number by the estimated number of hatchlings (198,181) enabled her to calculate the larval survivorship for that year at 0.14%. As mentioned in Chapter 8, when Vanessa processed these juvenile Crawfish Frogs, she gave each of them pond and year toe clips, so the next time we saw them we would know where they came from and their natal year.

Two years later, the first animals from the 2009 Nate's Pond cohort began showing up to breed. In 2011, Vanessa captured one male and three females at Nate's Pond, and one male at one of the other ponds she was monitoring at our study site. Vanessa's results corroborated Redmer's observation that Crawfish Frogs can breed in their second year.

The year 2011 produced the highest number of juveniles (3,122). By the next year, 2012, Vanessa had graduated with her master's degree and RMS had come onboard to begin her Ph.D. program. RMS took over from Vanessa, and in 2013 she discovered animals from the 2011 Nate's Pond cohort as follows. She captured 20 males and 10 females at Nate's Pond, and 6 males and 4 females at the other ponds we were monitoring at our study site (3 males and 1 female at Big Pond, 2 males and 2 females at Cattail Pond, 1 female at Erosion Control, and 1 male at Nate's Jr.).

In subsequent years RMS and her crew occasionally discovered Nate's Pond 2011 cohort animals as follows. In 2014 she captured 13 new males and

10 new females at Nate's Pond, and 8 males and 2 females at the other ponds she was monitoring (2 males and 1 female at Big Pond, 2 males at Cattail Pond, 3 males and 1 female at Erosion Control, and 1 male at New Pond).

In 2015 RMS captured three new males and four new females at Nate's Pond, and one male at Erosion Control.

In 2016 RMS captured just one new female at Nate's Pond.

In total, the 3,122 juveniles that metamorphosed in Nate's Pond in 2011 produced a total of 82 breeding adults (51 males, 31 females), a survivorship of 2.6%.

In contrast, survivorship among these adults in subsequent years was 80.5% from 2013 to 2014, 34.8% from 2014 to 2015, and 47.8% from 2015 to 2016. When RMS ran these analyses for other years at Nate's Pond, she got the same general trends, and her survivorship estimates by life history stage were as follows:

Egg: 98%
Larval: 1%
Juvenile: 3%
Adult: 70%

As an aside, Rochelle found that roughly two-thirds of all Crawfish Frog males bred in their second year, while roughly two-thirds of all Crawfish Frog females bred in their third year, confirming Redmer's conclusions.

From Vanessa's and RMS's data it is clear that the highest mortality in Crawfish Frogs occurs during the tadpole and juvenile stages. Compared with related ranid species, Crawfish Frogs exhibit a low larval and juvenile survivorship, but high adult survivorship. They can live for a long time. For example, Frog 53 (Corner Burn Chick) bred in Nate's Pond during all eight years of our study. The earliest she could have bred was as a two-year-old, making her at least 10 years old in 2016.* Given that most female Crawfish Frogs breed as three-year-olds, and that she may have begun breeding before our study began, there is every chance Corner Burn Chick was greater than 10 years old in 2016. Southern Leopard Frogs, a ranid that breeds at the same time as Crawfish Frogs (as mentioned in Chapter 9, they will also breed in the fall when conditions are favorable), and whose spring breeding tadpoles metamorphose at about the same time as Crawfish Frogs, have a maximum life span about half the age of Crawfish Frogs. If a Crawfish Frog can survive the bottleneck of high larval and juvenile mortality, it has a good chance of living a long life.

---

* J. Robb at Big Oaks National Wildlife Refuge in south-central Indiana has also documented a 10-year-old Crawfish Frog.

Boom, Bust, and Body Size **12**

After realizing that amphibian population declines were widespread and becoming a global problem, in the late 1980s scientists organized and began working to discover the nature of these threats and ways to blunt their effects. However, not all of the data pointing to amphibian declines at that time was robust. Preferring rigor, researchers at the Savannah River Ecology Lab, led by Joe Pechmann, urged caution in interpreting datasets suggesting declines by calling attention to the boom and bust cycles characteristic of many amphibian populations. These scientists described the results of their long-term study of breeding amphibians based on drift-fence data: "each species was common in some years but uncommon or absent in others" (i.e., species have periods of boom and periods of bust, sample in a bust period and what appears to be a decline might simply be a short-term response to local ecological conditions) (Pechmann et al. 1991). Later, David Green took this idea one step farther, and showed that the more variable the environment (for example breeding in unreliable ephemeral wetlands versus on a stable forest floor litter), the more amphibian populations fluctuate (boom and bust) (Green 2003). Further, Green demonstrated the amplitude of these boom–bust cycles correlated with the risk of population extirpation. Indeed, if a bust was severe enough, it would extirpate an amphibian population.

These types of assessments, and scores of others, have been made based on population size (typically, the number of breeding adults), which is a tangible, reliable technique. Drive into any small town on a blue highway and at the city limit you will see a sign with the name of the town and the number of people comprising its population.

But there is a second way to measure the strength of a population. At about the same time that amphibian declines were recognized as a global problem, researchers began using fitness metrics (size of individuals) to assess the success of amphibian populations (Semlitsch et al. 1988; Berven 1990; Wilbur 1997). Rather than simply counting the number of individuals in a population, fitness metrics use the positive relationship between female body size and fecundity (reproductive potential) to assess population vigor. In a nutshell, "larger [female] size directly affects fitness by ... increasing the numbers and size of eggs produced" (Berven 1990). In amphibians, size can also confer survival advantages through the speed of movements, endurance, resistance to desiccation, and predator avoidance.* As John-Alder and Morin state (John-Alder and Morin 1990): "size is a component of fitness." Imagine that small town sign at the city limit, but instead of the number of its inhabitants, it lists their average weight.

These two methods of evaluating amphibian populations—number of individuals versus size of individuals, representing values of quantity vs. quality—have mostly been treated separately in the scientific literature, yet there are practical considerations involved in knowing the dynamics of such a relationship. For example, coordinators of zoo-based head-start programs faced with limited budgets might wish to know which captive rearing approach will lead to the best outcome: raising large numbers of smaller animals (the population perspective), small numbers of larger animals (the fitness perspective), or some middle ground combination of number and size? (Stiles et al. in press). In natural populations, we can ask whether declines in numbers might be offset by an increase in size of the remaining individuals (or the reverse, whether gains in numbers of individuals are offset by smaller body sizes).

The quantitative approach, emphasizing the number of animals within a population, is the most widely used assessment, with high numbers naturally being desirable. From this perspective, high adult numbers lead to greater egg production, greater egg production leads to higher juvenile numbers, and higher juvenile numbers create even higher adult numbers, generating an upward population spiral. This more-is-better approach is successful up to a point. The wetlands supporting aquatic amphibians are bounded—they are not limitless, and as populations grow there comes a time when these basins can no longer support the number of larvae that have been produced, and a correction must occur. In ecology, such corrections fall under the category of density-dependent effects.† The seasonal and semi-permanent ponds that

---

* Summarized in Wells (2007), and Lannoo et al. (2017).

† An accurate assessment of larval density in aquatic amphibians is impossible to obtain, or even confidently estimate. While the general trend is for seasonal and semi-permanent wetlands to fill with snowmelt and rain water in the early spring, then gradually dry as the warm season progresses, this is far from predictable on a day-to-day, or year-to-year basis. Late-spring thunderstorms can refill drying basins, and in a wet spring following a dry winter, there may be more water present at the time of metamorphosis than there was three months earlier when eggs were laid. In general, drawdowns concentrate animals and increase density, while heavy predation reduces density. In many species, behavioral tendencies will increase effective density: bufonid tadpoles will school, and ranid tadpoles often aggregate in warmer shallows during sunny days, presumably to absorb heat and accelerate their development. Moreover, most density estimates fail to consider animal size. It is possible to fit 1,000 newly hatched Crawfish Frog tadpoles in a liter of water, but impossible to fit 1,000 pre-metamorphic tadpoles in the same volume. All of these uncertainties are in play when considering density-dependent effects on the animals experiencing them.

many species of amphibians use to breed are small (if they were large they'd be lakes and have predatory fish in them, which would exclude these amphibians). This means tadpole resources (algae, for example, which they feed on) are limited, and limited resources constrain growth. In the case of frog and toad tadpoles, density-dependent effects are caused by large numbers of conspecific tadpoles (within-species competition), or by large numbers of tadpoles of other species (across-species competition). The presence of predators—for example, carnivorous salamander larvae—on the other hand, should reduce tadpole density and relax these competitive effects.

To explore the impacts of competition and predation on Crawfish Frog tadpoles, Vanessa used field enclosure experiments to test four hypotheses (Stiles et al. in press). She predicted that:

1) within-species competition would produce smaller Crawfish Frog juveniles that took longer to metamorphose (Wilbur 1976, 1977a,b; Semlitsch and Caldwell 1982; Berven 1990);

2) across-species competition with Green Frog tadpoles or Southern Leopard Frog tadpoles would mimic intraspecific competition and produce smaller Crawfish Frog juveniles that took longer to metamorphose (Wilbur 1982; Alford and Wilbur 1985; Wilbur and Alford 1985; Werner 1992);

3) both within- and across-species competition would reduce survivorship to metamorphosis (Wilbur 1997; Semlitsch and Caldwell 1982; Smith 1983); and that

4) vertebrate (Smallmouth Salamander) predation on Crawfish Frog larvae would release competition pressure and produce relatively larger tadpoles that metamorphosed earlier (Wilbur 1972; Wilbur et al. 1983; Fauth 1990; Sredl and Collins 1992; Holbrook and Petranka 2004).

The data Vanessa gathered were consistent with most, but not all of her hypotheses. As she predicted, within-species competition reduced Crawfish Frog tadpole growth (both length and weight) and lengthened the time to metamorphosis (from an average of 56 to 61 days). She found a similar response with across-species competition. However, contrary to Vanessa's hypothesis, neither form of competition affected Crawfish Frog tadpole survivorship to metamorphosis. Apparently, competition does not kill Crawfish Frog tadpoles the way it can kill tadpoles in other species.

Vanessa also predicted that predation by Smallmouth Salamanders would severely reduce the survivorship of Crawfish Frog tadpoles, and she was right. More importantly, however, she was correct in her hypotheses that the density reduction of Crawfish Frog tadpoles caused by salamander predation would produce larger (longer and heavier) tadpoles that metamorphosed quicker than tadpoles raised at the same initial density without predators.

Following Vanessa's graduation in 2011, we continued her 2009–2011 field studies through the 2012–2016 breeding seasons, and used these collective data to determine whether the results from her enclosure study would translate into patterns detectable in the field. Indeed, they did. The initial density of Crawfish Frog larvae at Nate's Pond (dependent in part on the number of eggs laid that year by breeding females) correlated with the number of juveniles produced, while negatively affecting juvenile size (both length and weight) and days to metamorphosis (Stiles et al. in press).

The story doesn't end here. Researchers have shown that juvenile size and time to metamorphosis can carry-over to affect adult size—in essence, big newly metamorphosed juveniles retain their size advantage and become big breeding adults.* To quote Berven (1990): "larger juveniles and those metamorphosing early enjoy higher survival, earlier age at first reproduction, and are larger as adults." Berven then notes that larger females increase fitness by increasing the numbers and size of eggs they produce. Wilbur summarized the effects of this density dependence in aquatic amphibian larvae as follows (Wilbur 1997, p. 2287),

> Size at metamorphosis is an exponentially decreasing function of initial density of the population. At low density, many individuals metamorphose at a large size; at high densities most individuals metamorphose at what appears to be the minimum size threshold for successful metamorphosis. Survival is also an exponentially decreasing function of initial density, in part because as density is increased, reduced growth rate leads to a decreasing probability that an individual will obtain the minimum size threshold for metamorphosis before the pond dries or freezes.† These carry-over effects can also influence post-metamorphic survival (Goater 1994; Pechmann 1994; Beck and Congdon 2000; Chelgren et al. 2006), activity patterns (Yagi and Green 2018), and locomotion (John-Alder and Morin 1990; Alverez and Nicieza 2002; Cabrera-Guzmán et al. 2013).

---

* Semlitsch et al. (1988), Wilbur (1997), Beck and Congdon (2000), and Morey and Reznick (2001), a phenomenon called carry-over effects (Pechenik et al. 1998, Van Allen et al. 2010, Harrison et al. 2011, Earl and Semlitsch 2013, and O'Connor et al. 2014).

† Despite Wilbur's observation that size at metamorphosis is an exponentially decreasing function of *initial density* of the population, most field studies of density-dependent effects in amphibians with complex life histories use cohort size (*final density*) as the proxy for density. In fact, during years with wetland drawdowns and light predation, it is likely that the time immediately prior to metamorphosis is the time of highest tadpole density, but this may not always be true, and other proxies for density, such as number of breeding females and estimated number of eggs deposited—which assess densities earlier in the larval period—might serve just as well (and be consistent with Wilbur).

Building on Vanessa's reptarium study and our field data on the inverse relationships between tadpole number and juvenile size, we asked whether we would find evidence of density-dependent carry-over effects between juveniles and first-time (2-year-old*) breeding adults from these cohorts (Stiles et al. in press). Figure 12.1 summarizes these results as follows. Figure 12.1A: juvenile length (SVL) is inversely related to cohort size (density). There is slop in these data; 2009 was an outlier, but there is always much more variance in field data than experimental data, because field data do not control for all of the various biotic and abiotic influences on populations. Figure 12.1C: juvenile mass (grams) is also inversely related to cohort size. This is not surprising; in juvenile Crawfish Frogs, length and weight co-vary. Figure 12.1E: time to metamorphosis is positively correlated with cohort size. Each of these three results is consistent with Vanessa's enclosure study, and our overall understanding of density effects on tadpole growth and development. Figures 12.1B and 12.1D: juvenile cohort size is also related to adult length and weight. What is curious is, comparing Figures 12.1B and D to Figures 12.1A and B, the relationship between cohort size and Crawfish Frog size is stronger in adults than it is in juveniles. This suggests that whatever was holding back newly metamorphosed juveniles in the 2009 cohort was compensated for before these animals reached adulthood. Finally, Figure 12.1F: survivorship from metamorphosis to first breeding was independent of cohort size. This is the one result that contradicts prevailing thinking on fitness effects. High densities do not affect survivorship in either Crawfish Frog tadpoles or juveniles, as long as tadpoles reach a developmental stage advanced enough to permit metamorphosis.

Considered from a breeding adult perspective, the results we describe represent a specialized subset of our dataset—the four years when we could follow juveniles through to their first breeding. However, we have four more years of adult breeding data that add to this story. If we look at the relationship between the size and number of breeding female Crawfish Frogs, we see a startling relationship—the more females we have in the population, the smaller, on average, they tend to be, and vice-versa (Figure 12.2). Density dependence does not explain this relationship, because adult Crawfish Frogs live in isolation, and their density (on the order of 100 frogs per 2.5 kilometers²) does not tax their invertebrate prey base. Our video recordings show that the radius of feeding of a Crawfish Frog on its feeding platform is, at most, 45 centimeters (18 inches), which is an area of 0.64 meters². The area of a circle with a radius of 1 kilometer is 3.14 kilometers², which means that if there were enough habitable crayfish burrows, 1 square kilometer could theoretically accommodate

---

* When assessing adult size across cohorts, we restricted our analysis to include only data from 2-year-old, first-time breeding animals (i.e., to avoid confounding our results by mixing size data from these animals and larger 3-year-old animals).

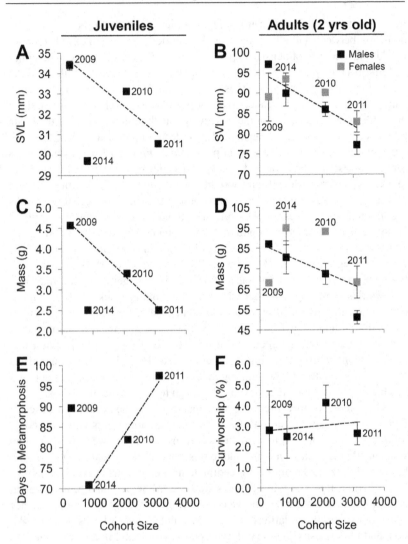

FIGURE 12.1   Field data showing the effects of juvenile density on juvenile size (A, C), time to metamorphosis (E), breeding adult size (B, D), and survivorship (F). Lines reflect regression calculations. While suggestive, none of these trends is significant except for the adult male SVL data shown in B. This image is used with the permission of Leo Smith and the American Society of Ichthyologists and Herpetologists.

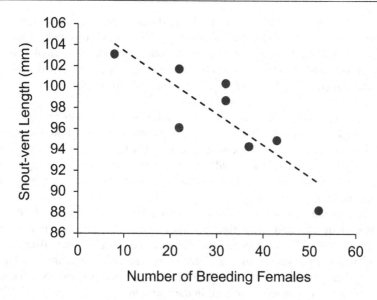

FIGURE 12.2 The relationship between the number of breeding females (x-axis) and their average size (SVL) by year is remarkably linear.

the feeding radii of 4,906,250 Crawfish Frogs. We do not think burrow-dwelling Crawfish Frog adults experience many density-dependent effects; indeed, few papers address density-dependent effects in terrestrial amphibians.

In fact, the only reasonable explanation is that the relationship between juvenile fitness metrics and adult fitness metrics that we observed during the four years that we could explore this linkage, holds for the four years when we could not. During the years of our study, the Crawfish Frog population at Nate's Pond appeared to be primarily driven by density-dependent effects in the larval stage.

Now, let's return to the concept of amphibian boom and bust cycles. Contemplating the data presented in Figure 12.2, we see that a "boom" in the context of number of individuals is a "bust" in terms of fitness metrics, and a "bust" in the context of number of individuals is a "boom" in terms of fitness.

It's time to re-examine fitness and carry-over effects by asking the question, under what environmental conditions do you find large numbers of small Crawfish Frog tadpoles (which in our dataset produced many, small Crawfish Frog adults) or, conversely, under what environmental conditions do you find small numbers of large tadpoles (which in our dataset produced few, large Crawfish Frog adults)? There are two scenarios we can imagine.

The first is a consequence of considering fitness as a driver of healthy amphibian populations. It goes like this. "Larger female body size (higher fitness) directly affects fitness by increasing the numbers and size of eggs produced" (Berven 1990). The number of eggs produced leads to a higher tadpole density, which leads to increased competition, which leads to smaller juveniles, which leads to smaller adults (reduced fitness). In essence, increased fitness through density-dependent effects eventually leads to smaller body sizes, which in turn leads to decreased fitness.

Now, carry this idea through. The smaller breeding adults produced by these density-dependent effects lay fewer eggs, which leads to fewer tadpoles, which leads to relaxed competitive effects. This reduced competition then reverses the fitness trend suggested in the previous paragraph, producing larger tadpoles, which develop into larger juveniles, which produce larger breeding adults and increased fitness.

To reiterate, the ultimate consequences of increased fitness effects are decreased fitness, and the ultimate consequences of decreased fitness are increased fitness. Thus, populations alternate their fitness states over time, with "numbers of individuals" and "size of individuals" interacting in a form of sinusoidal pattern (Figure 12.3). The odd thing about this scenario is that we have never seen it formally outlined in the literature.

The second driver of this process is external and is tied to density-independent environmental effects such as drought, increased predator densities, and disease. Each of these factors can reduce the numbers of Crawfish Frog tadpoles independent of their density. For example, in 2012, Nate's Pond dried before any Crawfish Frog tadpoles could reach metamorphosis, and no juveniles were produced. (A quick word about droughts. The presence of wet and dry years is not a random occurrence but rather is tied to the drought or hydrologic cycle, which in turn is tied to sunspot activity (Peplow 2004; Solanki et al. 2004). During the twentieth century, drought cycles occurred on an average of once every 10.7 years (Berger 2010), a pattern that until recently has been remarkably consistent.)

In the spring of 2013, large numbers of carnivorous Marbled Salamander larvae were present in Nate's Pond. We watched as these predators sat on Crawfish Frog egg masses and snapped at newly hatched tadpoles as they wriggled out of their egg casings. Later that year, only eight Crawfish Frog juveniles emerged from Nate's Pond.

Diseases such as chytrid fungus (*Batrachochytrium dendrobatidis*) and *Ranavirus* can also affect tadpole populations. Both are present at our study site, but only *Ranavirus* has proven to be a tadpole killer (Chapter 14).

Roughly one-third of female Crawfish Frogs reach sexual maturity in two years; the remaining two-thirds mature in three. That means that there will be a two- to three-year lag between environmental effects experienced by Crawfish Frog tadpoles, whether they be density-dependent or independent, and their impacts on the population. During years where there has been little

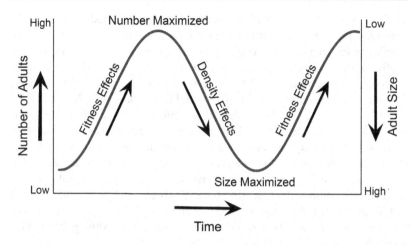

FIGURE 12.3 The proposed relationships over time between the number of breeding adult Crawfish Frogs and their size (fitness). Our data suggest an inverse relationship between these variables that oscillates over time, as follows. When adult numbers are maximized, the population experiences increased density effects on the relatively numerous offspring they produce. The smaller breeding adults produced lay fewer eggs, which leads to fewer tadpoles, which leads to relaxed density effects. This reduced competition then reverses the fitness trend, producing larger tadpoles, which develop into larger juveniles, which produce larger breeding adults and increased fitness. Thus, populations vary their fitness states over time, with "numbers of individuals" and "size of individuals" alternating in a form of sinusoidal pattern. External factors such as drought, increased predator densities, and disease may also drive this pattern by reducing the numbers of Crawfish Frog tadpoles independent of their initial density.

recruitment, the adult demographic of the population can be summarized as "fewer, older, and larger"—a bust year with each adult having a high potential fitness. In our dataset 2010 and 2012 were such years. In contrast, following years where there is high recruitment, the population is best summarized as "many, younger, and smaller"—a boom year with each adult having low potential fitness. In our dataset, 2011, 2013, and 2014 were such years.

A case study in tolerances. If the average estimated number of Crawfish Frog eggs laid each year at Nate's Pond (179,149) represents the reproductive effort required to maintain this population, we can ask how many of the largest female frogs in this population would be required to sustain it (minimum population size), and how many of the smallest females in this population would be required to sustain it?

Using the equation Clutch Size $= -10,974.3 + 172.4 \times SVL$ (see Chapter 11), we calculated the largest breeding female in our sample (115 mm SVL) would lay 8,675 eggs, while the smallest breeding female in our study (70 mm SVL) would lay 1,072 eggs. Dividing these numbers into 179,149 suggests 21 large females or 168 small females could sustain the population—numbers that in a relative sense would certainly qualify as bust or boom as judged from a simple headcount. Yet, considering fitness effects, any number of breeding females within this range could sustain the Crawfish Frog population at Nate's Pond.

Two loose ends. Wilbur explained the effects of high density on survivorship as follows: "as density is increased, reduced growth rate leads to a decreasing probability that an individual will obtain the minimum size threshold for metamorphosis before the pond dries or freezes" (Wilbur 1997). Yet, high larval densities do not affect survivorship in all species. Unlike several other studies (Wilbur 1997; Semlitsch and Caldwell 1982; Smith 1983; Alford and Wilbur 1985), Altwegg, working on the Pool Frog, *Pelophylax* (*Rana*) *lessonae*, found no density-dependent effects on survival (Altwegg 2003). Our results most resemble Altwegg's.

Drying pond conditions can also accelerate metamorphosis (Gomez-Mestre et al. 2013). As Wilbur and Collins point out (Wilbur and Collins 1973), species that inhabit uncertain environments such as seasonal and semi-permanent breeding ponds will have a wide range of possible sizes at metamorphosis, while species that inhabit more predictable environments will have a narrower range. In 2013 a population of Crawfish Frogs on private property (Ronk's Farm—likely the source population of HFWA-W Crawfish Frogs three decades ago), which bred at the same time as nearby (5 kilometers) HFWA-W Crawfish Frogs, produced larvae in shallow prairie swales that metamorphosed a full month earlier and a centimeter shorter (2.5 millimeters SVL; n=51) than HFWA-W animals. We suggest that species that have evolved wider ranges of sizes at metamorphosis have, within limits, the option to metamorphose earlier and smaller in the face of drying pond conditions. Therefore, these larvae will be more likely to survive desiccating conditions than frogs inhabiting more permanent aquatic systems. As mentioned earlier, Busby and Brecheisen offered that having arisen on the Great Plains, Crawfish Frogs evolved to breed in "buffalo wallows" (Busby and Brecheisen 1997), which are seasonal to semi-permanent bodies of freshwater.

There is one final consideration regarding the timing of metamorphosis (which determines how we measure the length of the larval stage) that has never been fully addressed, and which certainly contributes noise to the datasets being considered here. Some amphibian species produce larvae that metamorphose synchronously, often in response to an environmental trigger such as nighttime rain. Arnold and Wassersug proposed predator satiation to explain this phenomenon (Arnold and Wassersug 1978). Their hypothesis states that the probability of survival is increased when prey (in this case, newly metamorphosed juvenile frogs) metamorphose synchronously in order to overwhelm predators (such as snakes waiting in ambush along pond margins). Predators eat their fill, but because of the short-term, synchronous availability of prey, the window of opportunity to feed is small and fewer prey are taken. Indeed, newly metamorphosed Crawfish Frogs emerge following rains, which causes saltatory spurts of dispersing juveniles (Kinney 2011; Lannoo et al. 2017; Stiles et al. in press).

Now, if metamorphosis is defined by scientists not when animals actually metamorphose, but when juveniles emerge and appear in drift-fence buckets (technically, post-metamorphic dispersal), then metamorphosis as defined by these scientists is not some continuous, normally distributed function related to the completion of larval development, but rather the saltatory response of post-metamorphic animals to periodic nighttime rains—an environmental condition permitting distant dispersal across terrestrial landscapes hostile to animals prone to desiccation (Lannoo et al. 2017). It is possible that true metamorphosis occurs days, if not longer, before dispersing juveniles begin appearing in pitfall traps (Stiles et al. in press).

Where Do We Go from Here? 13

While we are confident that adult longevity in Crawfish Frogs is dependent on the habitation of crayfish burrows, we haven't collected enough data to understand, with confidence, why larval and juvenile survivorship is so low. We have, however, given this question a lot of thought. Busby and Brecheisen (1997) suggested that Crawfish Frogs evolved to breed in bison wallows. (Is there any more endangered wetland type than bison wallows?) Absent vast, migrating bison herds, these shallow wetlands are no longer being created or maintained, and any species that has tied their fate to them is bound to be in trouble.

Even if Crawfish Frogs did not evolve to breed in bison wallows, modern agriculture and associated drainage efforts across the southern Midwest and Great Plains have decimated most isolated wetlands. This is especially true of the shallow, fishless, seasonal and semi-permanent wetlands that many pond-breeding amphibians require for breeding. One explanation for the low larval survivorship we observe in Crawfish Frogs might be that the remnant wetlands on today's landscape are serving as refuges and attracting a larger number of species than historical wetlands ever did, making competition with the larvae of other frog species and predation pressure by larval salamanders and aquatic insects more severe than they were historically. A second explanation is that our results are the product of a small and restricted sample size—that the high larval mortality in our mine-spoil wetlands doesn't reflect the overall situation for Crawfish Frogs. MJL has often quipped that we know an awful lot about the biology of Crawfish Frogs on our 740 ha study site, and next to nothing about the Crawfish Frog population 3 kilometers away.

So how do Crawfish Frogs use the wetlands that are available to them? Our examination of Crawfish Frog genetics (Nunziata et al. 2013) suggested that while most Crawfish Frogs are faithful to their natal wetlands, some frogs wander. Since both adults and juveniles could be shifting wetlands, we wanted to understand the nature of Crawfish Frog movements across their prairie landscape. To do this, we tracked the wetlands and years where previously marked adults and juveniles bred (Lannoo et al. 2017).

Keeping track of adult frogs was simple. Every time we captured one, we noted its pit tag number and location. Of the 255 Crawfish Frogs we captured that had bred in at least two years, 57 (22%) of these animals chose to breed in two different wetlands.* Among these animals, 15 that first bred at Nate's Pond subsequently bred in other wetlands (9 at Big, 3 at Erosion Control, 2 at Cattail, and 1 at New). Fourteen animals that bred at Big Pond were subsequently captured at other wetlands, including 3 at Nate's Pond. Twenty-three animals that bred at Cattail Pond subsequently bred at nearby Big Pond.

At Nate's Pond, from 2012–2016, we captured and processed a total of 271 breeding adult Crawfish Frogs. Of these, 137 (50.6%) had toe clips indicating they were produced at Nate's and returned to breed. The remaining 134 adults (49.4%) were unmarked and therefore came from other wetlands.† In essence, half of the adults that entered Nate's Pond to breed were produced there; the other half were produced elsewhere (most likely at Big Pond).

Juveniles were another story. We had two ways of assessing the results of juvenile dispersal—animals dispersing from Nate's Pond to other wetlands, and animals dispersing from other wetlands to Nate's Pond. To determine juvenile dispersal from Nate's Pond, at each of the other five breeding wetlands (Cattail, Big, New, Erosion Control, and Nate's Jr.), we noted which breeding adults had been marked as newly metamorphosed juveniles with Nate's Pond toe clips. To determine dispersal into Nate's Pond, we noted first-time breeding animals without toe clips (we would have also counted animals with Cattail Pond toe clips, but we never recaptured any).

Of the 196 juveniles produced at Nate's Pond that we subsequently captured as breeding adults, 141 (71.9%) returned to Nate's Pond to breed; the remaining 55 juveniles (28.1%) dispersed. All five of the other breeding wetlands attracted Nate's Pond juveniles. Cattail and Big Ponds received the most dispersers (20 and 19 [10.2% and 9.7%], respectively). The greatest straight-line dispersal distance we documented was 1.35 km between Nate's and Nate's Jr. This distance was farther than the greatest distance we found between burrow site and breeding wetland (Frog 780) of 1.2 km (Heemeyer and Lannoo 2012). There were no predictable patterns to this dispersal. For example, the pond with by far the largest breeding population (Big Pond) received only the second-highest number of dispersing Nate's Pond adults, while the most distant pond (Nate's Jr.) received only the second-lowest number.

The direction and strength of juvenile dispersion differed from the direction and strength of adult migration. Presumably, this reflected the difference

---

* We say "bred" here, but realize that while nearly all females who entered wetlands during our study bred, not all males, especially small males, get the opportunity to breed.

† Because Crawfish Frogs take up to three years to reach sexual maturity, after 2012, the third year of our study, we assumed unmarked animals captured at Nate's Pond did not originate at Nate's Pond.

between dispersal movements (scattered, to places never experienced by these naïve newly metamorphosed juveniles) and migration movements (directed, by time-tested adults vectoring back to their primary burrows). In a uniform upland habitat, migration directions would mirror dispersal directions. But, of course, few habitats are uniform, even if they appear to us to be, and the differences between migration and dispersal directions likely reflect relative amounts of quality upland habitat (high crayfish burrow densities). They could also reflect other strong ecological relationships, such as differences in predator densities.

Classic metapopulation theory holds that movements among populations provide advantages such as repopulating habitats, avoiding inbreeding, and maintaining densities (Hanski 1998). Most metapopulation models assume dispersal occurs because individuals wish to avoid competition (Pillai et al. 2012). At our study site, the only resource we can imagine Crawfish Frog juveniles competing for is abandoned crayfish burrows. But as it turns out, juvenile Crawfish Frogs may not have the same affinity for abandoned crayfish burrows that we found in adults.

13.1 BORN TO RUN

In 2011, we recruited our colleague Mike Sisson to assist us with tracking newly metamorphosed juveniles using standard radiotelemetry techniques. Mike attached belted radio-transmitters* on 28 newly metamorphosed juveniles (Figure 13.1).[†] He released telemetered juveniles at a point near Nate's Pond, their natal wetland, and as they dispersed he recorded their locations. Mike measured distances between positions, and used these values to determine mean daily distance traveled, maximum daily distance, range of daily distances, farthest straight-line distance, and total distance moved during his study.

In 2015, Jonathan Swan constructed artificial burrows with dimensions based on burrows occupied by juvenile Crawfish Frogs.[‡] Jonathan augered 78 burrows within a 10-ha area. He obtained juvenile Crawfish Frogs from a cohort of RMS's head-started tadpoles reared at the Detroit Zoological Society (detailed in Chapters 14 and 16) (Stiles et al. 2016c). Jonathan released frogs at burrow entrances and monitored the fate of these animals using wildlife

---

* 0.48 g, BD-2N; Holohil Systems Ltd., Carp, Ontario, Canada.

† Plus three additional animals fitted with transmitters that generated no data.

‡ As follows: 50 mm in diameter, dug at a 45° angle, 1.2 m deep.

FIGURE 13.1   A juvenile Crawfish Frog sporting a belted radio-transmitter. We painted these transmitters brown to make them less conspicuous to predators.

cameras programmed to take photographs at one-minute intervals, around-the-clock. He followed a total of 149 juveniles from 24 June to 1 August and supplemented his wildlife camera data with radiotelemetry. He copied the techniques of, and equipment from, Mike Sisson's telemetry study, and tracked 12 juveniles from 28 June to 11 August, 2015.

Our previous drift-fence studies had suggested that unlike post-metamorphic Crawfish Frog adults, which leave breeding wetlands and vector in the direction of their primary burrows, newly metamorphosed Crawfish Frog juveniles dispersed from Nate's Pond in all directions; a pattern that, in another field, might be called dynamic disassembly. This result was confirmed by Sisson and Swan's telemetry studies. Telemetered juveniles released at a set point, whether near Nate's Pond or at an artificial burrow, dispersed without obvious directional preference (Figure 13.2). Sisson and Swan found that newly metamorphosed Crawfish Frogs will move an average of about 30 meters per day (35 meters in Sisson's 2011 study; 27 meters in Swan's 2015 study), although one juvenile moved a distance of 297 meters over the course of 24 hours. Sisson and Swan also found that as Crawfish Frog juveniles dispersed they generally vectored in a straight line (Figure 13.2), but tended not to follow drainages, paths, treelines, fence rows, or other natural or manmade features that might reduce resistance to movement or serve as obvious landmarks or guides to future breeding migrations (Lannoo et al. 2017).

Of the 149 juvenile Crawfish Frogs Swan released at artificial burrows, most left almost immediately (Figure 13.3); on average, they stayed at burrows

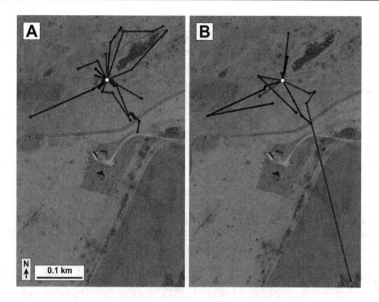

FIGURE 13.2 The results of Mike Sisson's radiotracking study following juvenile Crawfish Frog releases at the southern edge of Nate's Pond. Results are split into two figures for clarity of presentation. Note how frogs radiated in all directions (except through the pond). They tended to disperse in straight lines and did not appear to be following natural landscape features. This image was originally published in Lannoo et al. (2017) and is used with the permission of Leo Smith and the American Society of Ichthyologists and Herpetologists.

only two or three days (average = 2.7 days). However, several frogs remained at burrows for much longer, and one frog stayed 41 days. Among the frogs that remained, ten were preyed upon, usually by snakes or birds (Lannoo et al. 2017).

We made three observations that shed light on the behavior of dispersing newly metamorphosed Crawfish Frog juveniles. First, juveniles chose crayfish burrows with a diameter similar to those used by adults for their primary burrows, even though a large number of smaller-diameter burrows were available and presumably could have been used. Second, juvenile Crawfish Frogs achieve little reprieve from predators by occupying these relatively large-diameter crayfish burrows (Figures 13.4, 13.5). As noted above, while Crawfish Frog adults use crayfish burrows to protect them from predators as well as buffer environmental extremes, juveniles apparently use crayfish burrows only to buffer environmental extremes. And while post-breeding adult Crawfish Frogs rarely venture farther than one or two frog lengths from their burrow entrance, and will form a "feeding platform" as they trample and shade

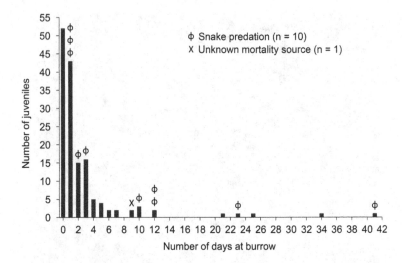

FIGURE 13.3 Retention of juvenile Crawfish Frogs released at artificial burrows at our study site from 24 June–1 August, 2015, as determined by wildlife cameras set to record at one-minute intervals. Average juvenile retention time was between two and three days. Of the 149 juveniles followed, 10 were either observed being eaten by snakes or disappeared shortly after snakes appeared on camera. This image was originally published in Lannoo et al. (2017) and is used with the permission of Leo Smith and the American Society of Ichthyologists and Herpetologists.

FIGURE 13.4 Juvenile Crawfish Frog in an abandoned crayfish burrow (compare this image to the adult shown in Figure 1.3). You can see the relatively large size of the burrow relative to the size of the frog. As far as we know, Crawfish Frog juveniles do not display the defensive tactics of adults (inflating their body against the burrow wall, lowering their head, etc.), and even if they did, these tactics could not work in the large burrows these animals tend to occupy. Photo by Andrew Hoffman and used with his permission.

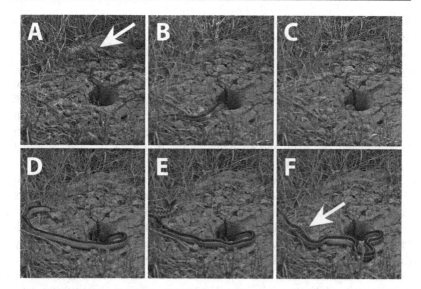

FIGURE 13.5 Not only do Crawfish Frog juveniles occupy large burrows, they tend to range from them. We initially followed this juvenile for a day and a half, as it entered and exited its burrow, and ranged back and forth between the burrow clearing and surrounding vegetation. Then, on 10 August, at 5:52 PM, it ranged south from the clearing into the surrounding vegetation (A, arrow). Forty-eight minutes later, a Gartersnake slipped into its burrow (B), and almost immediately turned around to position itself with its head on the burrow rim, facing west (C). It maintained that posture for the next 48 minutes, until the Crawfish Frog juvenile returned. The snake then grabbed the frog headfirst (D), and ingested it (E, F, arrow).

the vegetation where they position themselves, juvenile Crawfish Frogs range widely from their burrow entrance, putting themselves in a position to have little chance to reach their burrow when a predator attacks. Further, because juveniles naturally tend to occupy adult-sized burrows, they cannot inflate their bodies against the walls and lower their heads to dissuade predators, as adults will. Even if they did, their pointy snouts might work against them.

A curious aside. Not only are crayfish burrows unsupportive in Crawfish Frog juvenile defense against predators, our wildlife cameras revealed one instance where a burrow contributed to the death of a juvenile Crawfish Frog (Figure 13.5). Our first photo showing Juvenile 25 at a burrow was on 9 August, 2015, at 7:53 AM. We followed this frog for the next day and a half, at one-minute intervals, as it entered and exited this burrow, and moved back and forth between the burrow clearing and surrounding vegetation. On 10 August, at 5:52 PM, it moved south from the clearing into the surrounding vegetation (Figure 13.5A, arrowhead). Forty-eight minutes later, a Gartersnake slipped into its burrow (Figure 13.5B), and almost immediately turned around to position itself with its head on the burrow rim, facing west (Figure 13.5C). It maintained that posture for the next 48 minutes, until the Crawfish Frog juvenile returned. The snake grabbed the frog headfirst (Figure 13.5D, the spotting pattern and yellow-orange thigh region confirm this is a Crawfish Frog, and the size indicates it's a juvenile), and ingested it (Figure 13.5E, F, arrow). What we do not know is whether Gartersnakes just happen by burrows and co-opt them as ambush spots, or whether this snake detected the presence of the Crawfish Frog, probably by smell, and after finding the frog gone, decided to sit and wait, and ambush it on the chance it returned. It did.

Either way—if Gartersnakes use any old burrow as an opportunity to ambush or if they target occupied burrows—burrows that conceal predators become lethal locations for juvenile Crawfish Frogs. We do not know how common this behavior is by Gartersnakes, but of the 149 juvenile Crawfish Frogs Jonathan released at burrows, 10 were known to have been eaten by snakes. Taking this thinking to its extreme, burrow occupancy by juvenile Crawfish Frogs may have the opposite effect on survivorship that burrow occupancy in adults does—in juveniles, burrow occupancy may foster mortality (Lannoo et al. 2017).

Our third observation that sheds light on the behavior of dispersing newly metamorphosed Crawfish Frog juveniles, mentioned above, is that once newly metamorphosed Crawfish Frog juveniles locate and occupy a burrow, their overwhelming response is to abandon it in order to continue dispersing (Figure 13.2). Juvenile Crawfish Frogs apparently have a strong drive to disperse, and not only would time spent in retreat sites slow down their rate of dispersal, given the size of the burrows they choose and their behavior at them, burrow habitation likely does not reduce the rate of predation, and may, in fact, enhance it.

If burrow habitation by juvenile Crawfish Frogs does not thwart predation, predation pressure becomes some function of encounter rate with predators. That is, absent some mechanism to deter predators once you have been detected (for example, burrow defensive behaviors in adults), the best defense is to avoid encountering predators in the first place. With this in mind, we have suggested that dispersal itself functions to deter predation in juvenile Crawfish Frogs (Lannoo et al. 2017). Steve Arnold and Richard Wassersug* noted the tendency of certain snake species, such as Gartersnakes (*Thamnophis* sp.), to be attracted to the margins of ponds during amphibian metamorphosis (in our experience, so are Banded Watersnakes[†] and Prairie Kingsnakes[‡]). Given the

---

* Richard Wassersug was MJL's Ph.D. advisor and remains a cherished friend.
† *Nerodia fasciata.*
‡ *Lampropeltis calligaster.*

number of predators drawn to wetland margins, Crawfish Frog juveniles would enjoy a substantial survival advantage by getting as far away from a wetland edge as quickly as possible.

Beyond the wetland margins, Crawfish Frogs will continue to decrease their encounter rate with predators by decreasing their density, and can only decrease their density by dispersing (of course, predation will also reduce density). The post-metamorphic tendency by juvenile Crawfish Frogs to disperse continues for a month or more (it must for frogs to be found >1 kilometer from their breeding wetland, given an average juvenile movement of ~30 meters/day). Dispersal creates an enormous density advantage for Crawfish Frog juveniles. Using the equation for the area of a circle (πr^2), let us say 1,000 metamorphosing frogs are concentrated in the center of a drying wetland. Assuming no mortality, after the first dispersal movement—a maximum distance of 55 m in 2015—juvenile Crawfish Frogs increase their area occupied to 9,503 m^2, or one frog every 9.5 m^2 (most juveniles will not move the maximum distance so will be distributed within this circle). After the second movement, again 55 m, their density becomes one frog every 38 m^2. After the third movement, their density becomes one frog every 86 m^2, and so on in a geometric progression. After a few weeks of such movements, at a distance of 500 m, juvenile Crawfish Frog density becomes one frog every 78,000 m^2. For these frogs, dispersal not only serves to increase connectivity among populations, it may also serve as an effective anti-predator behavior by reducing frog–predator encounter rates. Much in the way we feel the habitation of abandoned crayfish burrows by Crawfish Frog *adults* is driven by the need to deter predators once they have been detected, we feel dispersal and the resistance to burrow habitation in Crawfish Frog *juveniles* is driven by the need to avoid detection through reducing encounters with predators (Lannoo et al. 2017).

Sick to Death

14

Scientists now realize that a number of diseases have recently hit amphibian populations so hard and so fast that any thorough investigation of population declines must now consider the impacts of these infectious agents. With this in mind, in 2010, we began screening Crawfish Frogs for chytrid fungus (*Bd*; an abbreviation for *Batrachochytrium dendrobatidis*, the scientific name of this pathogen) and *Ranavirus* infections.

Chytrids are a group of fungi that live in moist soil and aquatic habitats. There are perhaps 1,000 species of chytrid fungus, worldwide. Infections from two of these species, *Bd* and *Bsal* (*B. salamandrivorans*), under the right environmental conditions, can produce the disease chytridiomycosis in amphibians (Whittaker and Vredenburg 2011; Pessier and Mendelson 2017). *Ranaviruses* represent a subset of iridoviruses that infect not only amphibians, but also reptiles, especially turtles, and fishes. Both chytridiomycosis and *Ranaviruses* are considered emerging infectious diseases in that they, or the conditions for their expression, are spreading—affecting species and areas where they were previously unknown. Many scientists consider chytridiomycosis to be the principal driver of global amphibian declines (Whittaker and Vredenburg 2011).

As we processed Crawfish Frogs at our drift-fences, we screened for *Bd* and *Ranavirus* by swabbing every breeding adult we encountered, as well as a subset of newly metamorphosed juveniles. We suspected these results would provide insights on our overall knowledge of amphibian disease ecology, because when occupying crayfish burrows, Crawfish Frogs live singly—in nearly total isolation from other Crawfish Frogs—which we thought might slow the spread of disease. On the flipside, we felt these burrows, insulated from environmental extremes, might permit fungal spores and viruses to persist, maybe even thrive, enabling the spread of disease. Indeed, our results suggest these two factors are in play, at least for *Bd* infections.

Working with the veterinary pathologist, Allan Pessier, we found that 27% of all Crawfish Frogs *entering* our breeding wetlands tested *Bd*-positive, while almost twice as many, 46%, of all Crawfish Frogs *exiting* these wetlands tested *Bd*-positive (12% of these *Bd*-positive animals would develop the disease chytridiomycosis and die) (Figure 14.1). In contrast, not a single adult frog Jen

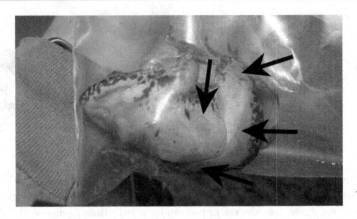

FIGURE 14.1   An adult female Crawfish Frog severely infected by chytrid fungus. While her deep-red lesions are not obvious in this black and white image, arrows indicate areas of greatest affect.

sampled while occupying upland crayfish burrows during the summer months tested *Bd*-positive. Similarly, Vanessa found that only 1% of the post-metamorphic juveniles she sampled tested positive for *Bd*.

In addition to examining *Bd* infection *frequencies*, we measured *Bd* infection *intensities*, measured as zoospore equivalents/swab.* Vance Vredenburg and his colleagues (Briggs et al. 2010) have suggested that a zoospore equivalent of approximately 10,000 triggers the disease chytridiomycosis and death in amphibians. In our dataset, zoospore equivalents/swab ranged from <1 to ~24,500. Consistent with Vredenburg's conclusion, the five frogs we sampled that died from chytridiomycosis had zoospore equivalents near or greater than 10,000 (Kinney et al. 2011).

The picture that emerges from our data indicates *Bd* infection rates in Crawfish Frog populations ratchet up from near zero during the summer to about 25% following overwintering; rates then continue to rise in wetlands, when frogs are concentrated and there is contact among frogs (male–male combat, male–female amplexus). *Bd* rates nearly double while frogs are in wetlands (Figure 14.2). The mortality we observed occurred just after breeding, during post-metamorphic migrations. Infections then wane during the summer.

The drivers of this cycle seem clear. During the late spring and early summer, following breeding, Crawfish Frogs clear *Bd* zoospores while basking on their hot, dry feeding platforms (we were able to assist Crawfish Frogs in shedding *Bd* by bringing severely infected animals back to our laboratory

---

* Today, field workers often identify *Bd* intensity using a PCR method that counts the number of 5.8S rDNA and ribosomal internal transcribed spacer (ITS) regions, termed ITS copies.

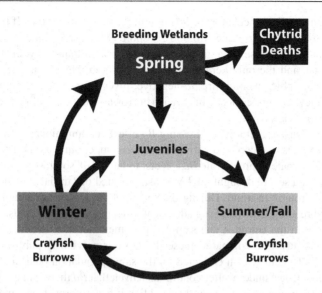

FIGURE 14.2 A simple model showing the patterns of chytrid fungus (*Bd*) gain and loss in Crawfish Frogs at our study site. Shades of gray indicate severity of infection. Following spring breeding, when *Bd* infections are highest, frogs shed the infection during the summer when their activity is centered at the burrow entrance and they bask. During the winter, frogs inhabit the water at the base of the burrow, where a subset (~25%) of animals then reacquires the infection. These infected animals then transmit *Bd* to their breeding wetland, and spread the infection (to roughly 50% of the population). Some animals (~3%) exiting wetlands develop the disease chytridiomycosis and die. Newly metamorphosed juvenile Crawfish Frogs were generally *Bd*-negative (1% infection rate). Juveniles are then either exposed to *Bd* while overwintering during the two years (males) or three years (females) prior to their first breeding attempts, or during their first breeding attempt. Once young Crawfish Frogs begin breeding, they follow the scenario outlined for breeding adults. This image was modified from a figure we published in Kinney et al. (2011).

and putting then in an incubator set at 30°C for 30 hours, which mimicked the effects of basking). The antimicrobial properties of Crawfish Frog skin-gland secretions (Ali et al. 2002) may facilitate shedding *Bd* infections. As they overwinter, about a quarter of Crawfish Frogs then reacquire *Bd* spores from the cool, moist walls of the burrow or the water in the chamber at its base. They then carry this infection with them to the breeding wetlands, and during the processes of male–male combat and male–female amplexus transfer zoospores, doubling the frequency of infection to around 50%. In addition to social factors, the cool spring weather and moist wetland conditions provide an excellent incubator for *Bd*, and infection intensities rise. If they rise enough to approach or exceed 10,000 zoospore equivalents, infected animals may succumb. Because *Bd* attacks keratin (the protein found in your hair and fingernails), and the only keratin tadpoles have is in their mouthparts, most post-metamorphic juveniles are *Bd*-negative. Juveniles may not be exposed to *Bd* until they take up residence in crayfish burrows, or until their first breeding, two or three years later.

The extensive literature addressing *Bd* effects on amphibians echoes our modest findings that ecologic variables such as temperature and moisture levels can exacerbate or ameliorate *Bd* effects. In late 2010, southwestern Indiana experienced a severe drought, and Vanessa repeated our initial *Bd* study with effects of drought in mind. During the spring of 2011, Vanessa sampled pre- and post-breeding Crawfish Frog adults entering and exiting wetlands, respectively; later in the summer, she sampled post-metamorphic juveniles as they exited wetlands. She compared these data to our previously published data from the 2010 field season collected on the same population (and in 19 cases on the same frogs) under wetter conditions. Given that *Bd* thrives in cool, high-humidity environments, we hypothesized that infection prevalence and intensity, as well as observed mortality due to chytridiomycosis, would be lower in 2011 than the levels we had observed in 2010.

Indeed, this was the case. In 2011, we analyzed 111 *Bd* samples from Crawfish Frog breeding adults: 62 samples from frogs entering wetlands, 49 from frogs exiting; 42 frogs were sampled both entering and exiting wetlands (Terrell et al. 2014a). As mentioned above, 19 frogs had been sampled previously. In addition, we swabbed 430 newly metamorphosed juvenile Crawfish Frogs, and from this sample, we randomly subsampled 101 swabs.

In our 2011 dataset, 13 of 62 (21%) samples from pre-breeding frogs tested positive, while 21 of 49 (43%) of post-breeding adults tested positive. These prevalence data were not statistically different from our 2010 data. Further, in 2011, we found no *Bd*-positive samples from the 101 post-metamorphic juveniles we swabbed. Again, this result was similar to data collected in 2010, when 1% of animals sampled were *Bd*-positive. An interesting aside: of the 19 Crawfish Frogs Vanessa swabbed in 2011 that she had previously swabbed in 2010, two animals were positive both years, seven were negative both years, six were negative in 2010 and positive in 2011, and four were positive in 2010 and negative in 2011.

We also analyzed the *Bd* infection intensity (zoospore equivalent/swab) data, and discovered an interesting disconnection. Up to that point, the scientific literature had indicated that the rate of *Bd* infection across members of a population (prevalence) was proportional to the intensity of infection within members of a population (intensity). That is, the higher the proportion of infected individuals, the more severe the infection was for each individual. Vanessa's data, however, showed that under drought conditions, this relationship between infection prevalence and intensity was uncoupled. While our prevalence data from 2011 was statistically identical to her prevalence data from 2010, our intensity data (zoospore equivalents/swab) during the drought year of 2011 showed a roughly 200-fold drop. Among pre-breeding adults *Bd* intensity fell from 482 zoospore equivalents in 2010, to 2.6 in 2011; among post-breeding adults, intensity dropped from 3,707 zoospore equivalents in 2010 to just 18.4 in 2011 (Figure 14.3). Furthermore, unlike in 2010, where five frogs died from chytridiomycosis, in 2011, we recorded no *Bd*-related frog deaths.

The constant prevalence rates we observed in both years suggest either that Crawfish Frogs are being exposed to *Bd* sources independent of ambient moisture, or that low-level infections below detection thresholds persist from year to year. These results also made us rethink the effects of drought, which is generally perceived to be disastrous for amphibians. Drought does, however, have occasional beneficial ecological effects for amphibians. Drawn-down seasonal and semi-permanent wetlands lose their fish and invertebrate fauna; absent these predators, tadpoles can thrive. To this ecologic benefit of drought, we suggested another—that drought can reduce *Bd* infection intensities, and therefore decrease the incidence of amphibian deaths due to chytridiomycosis.

In 2014, we discovered that in addition to *Bd*, *Ranavirus* was present at our study site (Stiles et al. 2016b). On 31 July, RMS and technicians from the Detroit Zoological Society returned almost 4,000 Crawfish Frog tadpoles they had raised from an egg mass collected at Cattail Pond three months earlier. The plan was for these tadpoles to complete their larval stage in submerged cages at HFWA-W until they reached metamorphosis, then RMS would release the newly metamorphosed juveniles to the area around Cattail Pond. Soon after placing her tadpoles in their cages, however, they began exhibiting abdominal and limb bloating, and hemorrhaging around their mouths and on their limbs. Some tadpoles developed skin ulcers. All of these sick animals quickly became lethargic and began dying. We suspected *Ranavirus* and sent frozen livers and kidneys from tadpole necropsies to Dr. Pessier, who confirmed that all samples were strongly positive for *Ranavirus* DNA.

Using the series of swabs we had previously collected and stored throughout the course of this study, we determined that *Ranavirus* had been present on the egg mass we'd collected at Cattail Pond, and that this pathogen had been transferred to tadpoles at hatching and persisted through their early development at the Detroit Zoo. Transport-related stress, as well as natural immunosuppression as animals approached metamorphosis, likely kindled this disease in infected tadpoles, and then spread quickly throughout the caged tadpole aggregations at HFWA-W. This was the only time in the eight years of our Crawfish Frog study we observed deaths due to *Ranavirus* at our study site.

Collectively, our field data suggested that both *Bd* and *Ranavirus* are always present at our study site, but that special conditions—cold, wet spring weather in the case of *Bd*; stress in infected, crowded animals in the case of *Ranavirus*—are necessary for these diseases to become fulminant and deadly.

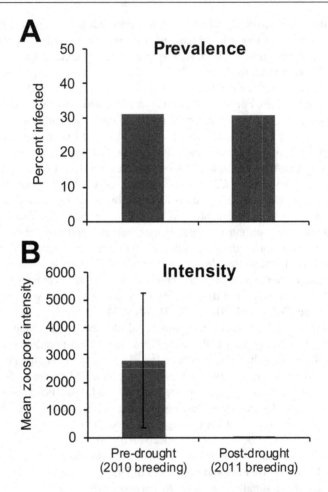

FIGURE 14.3 Comparisons of prevalence and intensity of positive samples (mean zoospore equivalents) of chytrid fungus (*Bd*) infection in breeding Crawfish Frog adults in the moderately wet year of 2010, and following an extreme drought in 2011. (A) *Bd* prevalences were not significantly different between 2010 and 2011. (B) In contrast, *Bd* intensities were substantially different between 2010 and 2011, with 2011 intensities roughly 400 times lower than in 2010. We suggest (Terrell et al. 2014a) that while Crawfish Frogs are being exposed and infected by *Bd* (likely in their burrows) similarly regardless of the environmental conditions, cool and wet conditions favor infections and can lead to the disease chytridiomycosis.

True Frogs Don't Sleep

15

As mentioned in Chapter 6, in 2009, we began using wildlife cameras to elucidate the activity patterns of Crawfish Frogs (Hoffman et al. 2010). It began by accident. Jen was tracking a frog, 460 (see below) who she never saw. Each day she would trek up the hill, her radiotracking receiver beeping 460's frequency, and watch his burrow entrance for signs of movement. She never saw any. Standing over his burrow, his telemetry signal came in strong, but after several weeks she began to wonder, was he alive and in his burrow, did he shed his transmitter and leave the burrow, or did he die in his burrow? She couldn't tell. We talked many times over several days about how to best answer this question. Our first thought was to dig him out, but doing that meant destroying the integrity of his burrow, and if he was alive the excavation would leave him homeless, and likely snake bit. We finally decided to purchase a wildlife camera and set up a surveillance on his burrow (Figure 15.1). It only took us a few days of scanning images to realize that 460 was almost always out of his burrow, except when Jen approached. This shy frog would jump in his burrow, apparently at the sound of her approach and maybe her beeping receiver, stay there while she took Kestrel® readings, then re-emerge shortly after she left. More importantly, however, as we flipped through the digital images, we discovered the camera followed his activities during the day (in color) and at night (using the black and white infrared feature). We then realized the potential of wildlife cameras to track these sedentary Crawfish Frogs for as long as they inhabited their primary burrows. Further, wildlife cameras have the advantage of not encumbering the animal with a radio-transmitter, plus they provide a visual record of what an animal was doing at the time of each activity assessment.

We then teamed up with Drs. Daryl Karns and Joe Robb, and their students and technicians at Big Oaks National Wildlife Refuge. Using a variety of video and still-motion technologies, we demonstrated that Crawfish Frogs exhibit around-the-clock (circumdiel) activity patterns, and do not appear to sleep or exhibit winter torpor (Hoffman et al. 2010).

In 2014, RMS took the lead on a follow-up study using newer, high-resolution (down to five-minute intervals) wildlife cameras deployed

FIGURE 15.1 Throughout our study we used wildlife cameras to monitor the activity of Crawfish Frogs at their primary burrows.

around-the-clock and throughout the year to monitor the behavior of Crawfish Frog adults and juveniles. This study lasted five years (with Nate doing much of the early grunt work, such as changing batteries and memory cards, downloading memory cards, and retrofitting cameras for the close-up photographs we required), and produced some fascinating results (Stiles et al. 2017). For one, although Crawfish Frogs were most active during the warm season, from April to October, we observed Crawfish Frogs up and outside their burrows every month of the year. More interesting, however, was that Crawfish Frogs were not uniformly active during the warm season. Peaks of highest activity occurred in May and September when frogs were active day and night. Early spring and late fall lulls were due to frogs being active during the day (diurnal), but not at night, when ambient temperatures dipped close to freezing. The summer lull was due to frogs being active at night (nocturnal), in order to avoid daytime hot and dry (desiccating) conditions. The moderate temperature and moisture conditions of the late spring and early fall seasons enabled Crawfish Frogs to be active around-the-clock.

Despite typically living alone in upland burrows and not being in social contact with any other Crawfish Frog, the activity patterns of individual frogs were uncannily synchronous, shifting from diurnal to circumdiel to nocturnal and back to circumdiel then diurnal as the year progressed, as if they were choreographed. Neither the sex nor the age of these frogs mattered; we found no differences between the activity patterns of adult males and females, or between adults and juveniles (Figures 15.2 and 15.3). Because these frogs live

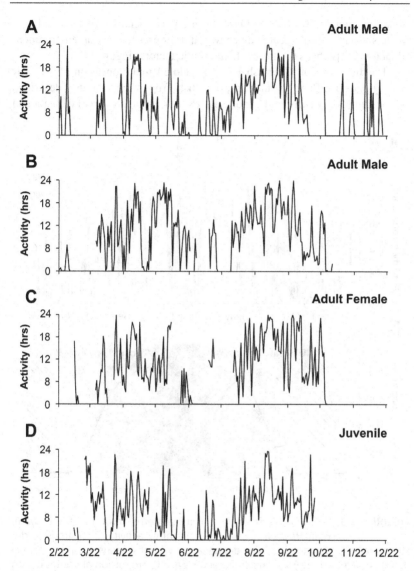

FIGURE 15.2 Daily activity of the four longest monitored Crawfish Frogs at our study site in 2012. (A) and (B) adult males (Frogs 44 and 26 [Romeo], respectively); (C) an adult female (Frog 53 [Corner Burn Chick]); and (D) a juvenile. We measured activity as the percent of images/day during which a frog was active (i.e., out of its burrow). Notice the similarity in activity patterns across sexes and life stages. This image was originally published in Stiles et al. (2017) and is used with permission of Bruce Bury and Herpetological Conservation and Biology.

alone, this synchrony could not have been due to social factors; it was likely due to similar physiological and behavioral responses to external environmental factors, or perhaps to internal homeostatic factors (Figure 15.3).

The literature on Crawfish Frogs is riddled with confusion about their activity patterns. Prior to our work, Crawfish Frogs were variously described as emerging "only early in the morning" (Smith 1950), nocturnal (Conant and

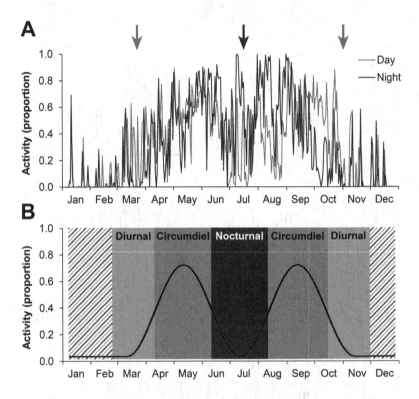

FIGURE 15.3 Crawfish Frogs shift activity patterns throughout the warm months of the year. Here, we show mean monthly activity of 17 frogs at our study site, sorted by total activity, diurnal activity (sunrise–sunset), and nocturnal activity (sunset–sunrise). We measured activity as the proportion of photos during which a frog was active on a given day, parsed into daytime or nighttime hours. We then averaged these data for all frogs by month. Activity peaks correspond to around-the-clock (circumdiel) activity in the late spring (May) and early fall (September). Early spring increases (March–April) in activity and late fall decreases (October) in activity are due to animals being primarily diurnal, likely due to cool nighttime temperatures. The summer dip in activity (centered in July) corresponds to decreased diurnal activity, likely due to the avoidance of mid-summer daytime hot and dry conditions. This image was modified from Stiles et al. (2017).

Collins 1998; Minton 2001; Parris and Redmer 2005), nocturnal following rains (Johnson 2000), or crepuscular (Thompson 1915). As we pointed out in 2010, these authors were not wrong, they were incompletely correct—they never realized they were observing only portions of a complex behavioral plasticity (Hoffman et al. 2010).

Everyone who knows frogs knows that they feed during the day—they are visual predators—in contrast they call and breed at night. Further, frogs in the family Ranidae do not sleep; they do not assume the classic anuran sleep posture of tucking their limbs against their trunks and closing their eyes (as do hylid treefrogs and bufonid toads), nor do they exhibit brain waves characteristic of sleep (Hobson 1967, see also Huntley et al. 1978). Hobson concludes that ranids such as Crawfish Frogs "rest without loss of vigilance" (Hobson 1967).

Our results suggest that Crawfish Frogs synchronously alter their activity patterns in response to varying environmental temperature and moisture conditions. This behavioral plasticity may provide them, and other ranid frogs, with a previously unrecognized resilience to ecological variations associated with climate change (McCarty 2001; Diffenbaugh et al. 2005; Parmesan 2006). Researchers have recently demonstrated that behavioral and morphological plasticity, as well as genetic adaptation, can offset some of the negative consequences of climate change (Reading 1998; Urban et al. 2013). We agree, and RMS has suggested that the behavioral plasticity that allows Crawfish Frogs to be active day and/or night gives them some capacity to be naturally resilient to the current effects of climate extremes.

The Frog That Gambled and Lost

16

Crawfish Frogs have lost at least a third of their geographic range since scientists began documenting their distribution, and probably a larger percentage of their historical numbers (since most populations have been reduced and many are just hanging on). For example in Indiana, Nate showed that Crawfish Frogs have experienced declines throughout most of their range (Engbrecht and Lannoo 2010; Engbrecht et al. 2013). He conducted surveys at nine historical sites and detected Crawfish Frogs at only one of them. Nate's data suggest Crawfish Frogs have been extirpated from Benton, Fountain, and Vermillion counties in the north, Vanderburgh and Warrick counties in the south, and Morgan and Monroe counties in the east. Robust populations (~300 animals) persist in only two areas: on our study site at HFWA-W in the southwest portion of the state, and at Big Oaks National Wildlife Refuge in the southeast. A third population cluster remains in Spencer County, in the south. Remaining animals are scattered among populations that are generally small and located on private lands in southwestern Indiana. Nate's data suggest there are fewer than 1,000 adult Crawfish Frogs remaining in Indiana.

It's the same story across the 13-state range of Crawfish Frogs. Of the 243 counties where Crawfish Frogs were known to occur, by 2016 they appeared to have been extirpated from at least 84, a 35% reduction. The situation is worse east of the Mississippi River. Of the 86 counties in the six states encompassing the range of Crawfish Frogs, they are thought to be extirpated from 52 of them, a 59% reduction.

Despite these losses, not all is lost. Crawfish Frogs are resilient and will establish populations at new sites when habitat becomes available and animals are close enough to colonize. One huge advantage Crawfish Frogs have over many other threatened and endangered species is that they can produce large numbers of offspring—Crawfish Frogs have no trouble making babies. It's the survivorship of these offspring during the tadpole and juvenile stages where the conservation challenges occur. We do not know how to improve the survivorship of Crawfish Frog juveniles—there is little we can do to manipulate conditions to make them favorable for these mobile, free-ranging animals. On the other hand, through head-starting initiatives, we found that we can greatly improve the survivorship of Crawfish Frog tadpoles. Survival rates of captive-reared Crawfish Frog tadpoles exceeded those of wild tadpoles by nearly a 100-fold. In the wild, an egg mass containing 5,000 eggs is required to produce 50 newly metamorphosed juveniles. In captivity, using the techniques that RMS and her colleagues at the Detroit Zoological Society developed, the same-sized egg mass can produce 4,500 juveniles (Stiles et al. 2016c). This is what's called hope.

Throughout the scientific literature, Crawfish Frogs have been called slow, clumsy, and stupid—characteristics that have been interpreted as producing the poor competitive performances of tadpoles and juveniles. There seems to be a strong tendency in humans to blame victims for the crimes committed against them—i.e., if they had only been smarter, more aware, and/or less naïve they could have avoided their misfortune. This type of thinking often underlies discussions of Crawfish Frog conservation and it goes something like this: "If Crawfish Frogs were not so slow, clumsy, and stupid they might muster enough wherewithal to pull themselves up by the bootstraps and save themselves from extinction. Crawfish Frogs should be blamed for their own declines." Further, Crawfish Frogs are big, and the human association between big and dumb (think Lennie in John Steinbeck's *Of Mice and Men*) does not help this innate bias.

The old-timers can be excused for thinking this way. Considering Crawfish Frog behaviors relative to other frogs, Crawfish Frogs *are* slow. When we were processing frogs at our drift-fences, we would remove a frog from a bucket, take a measurement, set the frog down beside us, record the measurement, pick the frog up, take another measurement, set the frog down, record, etc. This didn't work with Southern Leopard Frogs; we'd release them and they'd bound off. But remember, unlike Southern Leopard Frogs, who flee in response to predators, Crawfish Frogs hold their ground. When in a burrow, it is its non-movement that makes a Crawfish Frog able to consistently survive a snake attack (something no other North American frog can do). The frog Nate found in the belly of a Hognose Snake didn't die because she remained still with her body inflated and her head down, she died because she lifted her head—she moved.

When we began our Crawfish Frog project, we knew populations were in severe decline. We also knew they bred in the spring, in shallow, fishless wetlands, along with Spring Peepers, Chorus Frogs, Southern Leopard Frogs, Smallmouth Salamanders, and Tiger Salamanders. When considering research strategies, our thinking went like this. From among this guild of spring breeding amphibian species, only Crawfish Frogs were endangered. This suggested to us that their conservation challenges were due to terrestrial upland issues, not aquatic wetland issues. Indeed, we now feel this is the case, and we think we know why. Crawfish Frogs inhabiting crayfish burrows are immune from most environmental insults. We have observed them easily survive when their burrows have been mowed or burned over; they even appear to be immune to the effects of herbicide (glyphosate) applications. What they cannot survive, however, is the universal terrestrial disturbance—soil tillage.

Imagine a Southern Leopard Frog sitting near a Crawfish Frog at its burrow with an oversized modern agricultural tractor bearing down on them, pulling a moldboard plow rig. The Leopard Frog hops sideways three or four or eight times and is out of harm's way. The Crawfish Frog, on the other hand, dives 3 or 4 inches into its burrow, turns around, lowers its head, and puffs up. Who lives and who dies? In pre-settlement tallgrass prairie ecosystems, Crawfish Frogs compensated for high larval and juvenile mortalities by living long lives and reproducing many times. In contrast, in many of today's populations, adults are regularly being plowed up and killed. Add these adult losses to natural larval and juvenile mortality and it is no wonder Crawfish Frog populations are winking out.

The key to Crawfish Frog conservation, therefore, is to cease plowing (including the management technique of strip-disking) in areas where we wish to conserve them. Perfect. This seems like a straightforward and reasonable management recommendation. But since Crawfish Frog burrows can be over a kilometer from their breeding wetlands, following this advice means restricting agricultural activities within a 1-kilometer radius of Crawfish Frog breeding wetlands. For those of us used to thinking in Standard English measurements, that's an area encompassing a square mile. And a square mile taken out of agricultural production has seemed to society to be an awfully high price to pay to conserve a population of frogs people occasionally hear but never see.

If there is any behavior that Crawfish Frogs exhibit that should be held responsible for their declines, it is not their sluggishness, instead it is two activities related to using crayfish burrows as upland habitat, as follows. First, adopting a single home burrow they inhabit for up to five years; these "homes" tie Crawfish Frogs to a tiny (0.05 m^2) spot on Earth for half a decade, far longer than the average interval between plowing events in regions of the country now supporting row crop agriculture. Busby and Brecheisen observe "the largest remaining populations … occur where extensive tracts of prairie remain" (Busby and Brecheisen 1997, p. 216). Second, employing the strategy of diving into these burrows, quickly turning around, puffing up, and lowering their head in response to threats. This is a superb defense against snakes, a poor defense against agricultural tilling. As Thompson noted, "they are plowed out in numbers and the ground in that region is only plowed to a depth of about three inches" (Thompson 1915). Similarly, Hobart Smith and his colleagues note: "specimens were being plowed up in fields at a depth of about 6 inches" (Smith et al. 1948, p. 609).

There can be no doubt that the absence of plowing is the reason why remaining Crawfish Frog strongholds are along the eastern border of the Great Plains (eastern Kansas, Oklahoma, and Texas), where ranching predominates. It is also why the remaining two Crawfish Frog strongholds in Indiana are on large tracts of land set aside for wildlife (Big Oaks National Wildlife Refuge [federal], and HFWA-W [state]). Further proof comes from population estimates at our study site. Following the 2011 breeding season, in response to Jen's recommendations and after some high-level discussions, the property manager at HFWA-W imposed a no-plow policy. The result of this policy was a more than doubling of the Crawfish Frog population over the next four years (Figure 16.1).

Of course, local Crawfish Frog declines can be produced by other causes, such as wetland drainage eliminating breeding habitat, upland crayfish declines, and diseases such as those caused by fungal, viral, and bacterial pathogens. But given our experiences and our data, we feel the main driver of their declines has been agricultural tillage.

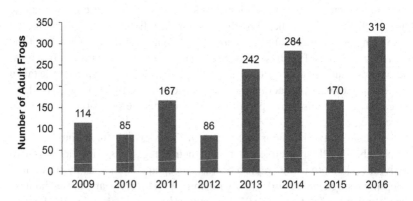

FIGURE 16.1 Total numbers of breeding Crawfish Frogs captured over the course of our study, by year. The low years 2010 and 2015 corresponded to pre-breeding season prescribed burns around Nate's Pond, which inhibited breeding migrations; the 2012 lull was due to a cold snap. What is instructive is to look at the years when these factors were not in play. Every year of our study without these influences (2009, 2011, 2013, 2014, and 2016), Crawfish Frog numbers increased, due in part, we believe, to implementing management suggestions such as reducing tillage in areas where Crawfish Frogs were inhabiting burrows.

Frog Tales 17

17.1 FROG 538—BAD LUCK

Frog 538 was a young, possibly 3-year-old female when she showed up gravid in Bucket 22 at Nate's Pond on 1 May, 2009; we initially called her May Day. Two behaviors distinguished her. First, she showed up late—two weeks later than the previous last female; second, she was in amplexus with a small male frog we came to call 401 (Figure 17.1). (In 2010, Frog 401 switched breeding ponds and migrated to Big Pond. Not long after, on 10 April, Jen found him in the belly of a Gartersnake.) Frog 538 was the first, and remains the only, female we discovered in amplexus in a bucket, and at the time we suspected that her tardiness may have been due to her ferrying 401, perhaps a long way. (We still don't know whether or not this is true, but at 5:00 AM on 3 April 2015 we found a male ferrying another male across land during their post-breeding migration, which proved Crawfish Frogs will carry other Crawfish Frogs during breeding migrations, and male–male ferrying seemed to us to be a much less probable occurrence than a pre-breeding male on female observation). At any rate, 538 bred late, and was accompanied by a beau. She laid her egg mass and exited Nate's Pond five days later, from Bucket 4. Jen implanted a transmitter and released her (Figure 17.2). Subsequently, 538 crossed a road and settled into an area about 250 m northwest of Nate's Pond. It took her a long time to find a burrow, and it's a good thing she finally did, because the DNR burned the area on 19 September. It was a hot burn and all vegetation was eliminated. A month later, on 30 October, coyotes tried to dig out her and a nearby male Crawfish Frog from their burrows, and while these attempts were unsuccessful, the excavation caused 538's burrow to collapse. Rather than spend the winter at the base of a meter-deep burrow, in the relative warmth below the frost line, she was forced to spend the winter in water and ice near the top of her burrow (Figure 17.3).

538 survived the winter; don't ask us how. But during the following spring she remained in her burrow long after Jen's other telemetered frogs had begun their pre-breeding migrations. She finally emerged on 28 March, muddy but

FIGURE 17.1   Male Frog 401 (top) and female Frog 538 in amplexus on 1 May, 2009.

FIGURE 17.2   Jen Heemeyer tracking Crawfish Frogs. This area of our study site had been prescribed burned the previous fall.

apparently healthy (Figure 17.4). Nine days later, on 6 April, she began her pre-breeding migration, but got only 100 meters or so and spent the night on her side of the road. The next night she crossed the road and made it about halfway to Nate's Pond. There, a predator nabbed her. The next morning Jen found her mangled body. It was tragic. We know we're supposed to be objective scientists, and we most certainly are when it comes to our collection and

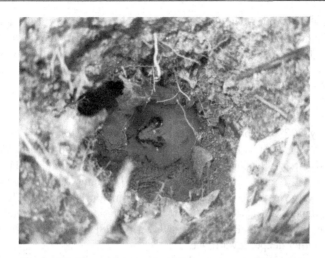

FIGURE 17.3 Frog 538 overwintering in her burrow after it had collapsed following a failed attempt by coyotes to dig her out.

FIGURE 17.4 The last photo of Frog 538 taken while she was alive. Notice her back is covered in mud, reflecting the difficult winter she had just experienced. The following night she was killed by an unknown predator.

analysis of data. But when observing animals like Frog 538 survive an entire winter against impossibly long odds, it's human nature to root for them; and we unapologetically pulled for Frog 538. We have no doubt that this boosterism for individual frogs comes from the same deep emotional source as our desire to save this species from extinction.

Here's how our field notes tracked Frog 538's fate:

6 May, 2009: Jen put transmitter in ♀ that came in 1 May.

30 October, 2009: Jen called to say ♂ and ♀ [538] in burn had burrows partially dug out by coyotes. Frogs are OK although burrow entrances are a mess. Are we leading [predators to occupied burrows]? We weren't there yesterday so think [our] scent would be old.

18 January, 2010: Holes in burrows in burn dug [up] by coyotes are really exposed. Big holes. ♀ with ice [538] had surface ice over another deeper layer of ice.

4 February, 2010: Jen showed us pictures of ♀ [538] in burn in winter. Three images—ice, water, frog.

3 March, 2010: ♀ [538] in burn near surface looking bad but alive. Burrow caved in? Short tunnel, 1 cm of water. Took pictures and put her back. Will check on her tomorrow.

5 March, 2010: [♀] 538 still alive in water in shallow (non-)burrow.

13 March, 2010: When we approached [538, we] could see frog in burrow—looked dead. Pulled out, was still alive. Burrow is shallow ~1 cm of water in burrow. Took pictures, swabbed [for chytrid fungus], and returned [her] to burrow.

24 March, 2010: last night [was] first full [Crawfish Frog] chorus. Almost all animals moved. 538 [had] moved 3–4 m toward Nate's [Pond] by 10:30 PM but [then] went back to her burrow.

25 March, 2010: 538 still in burrow.

28 March, 2010 [at] 1:00 AM: 538 out of her burrow—at entrance.

28 March, 2010 at 6:00 PM: 538 still at burrow—out—burrow flooded—the only frog w/radio still associated with burrow [i.e., had not at least begun its breeding migration].

3 April, 2010: Yesterday 538 moved to [a] new burrow, today [she] moved back.

7 April, 2010 AM: 538 was killed in wet seep on N(orth) side of mowed path [path was mowed previous fall by DNR to assist quail hunters]. Found this morning by Jen. Looked squished—viscera including eggs out of mouth. Skin lacerated on side of head.

8 April, 2010: 401 [the male 538 was found in amplexus with in 2009] eaten by a Gartersnake. Jen found, has carcass [she rolled the snake to make it regurgitate the frog].

17.2 FROG 460—PARTY BURROW

We've mentioned Frog 460 before, when Jen never saw him and had concerns about his health. In early May, 2009 we conjured up the idea of putting a wildlife camera on him, and learned he was simply shy. This frog was the only telemetered frog we followed that lived in a burrow with multiple openings. This may be the reason that in addition to Frog 460, our digital images captured a Gartersnake, small mammal (field mouse or vole), and a small crayfish also occupying his burrow, which we dubbed the party burrow (Figure 17.5).

17.3 FROG 53—CORNER BURN CHICK

This shy frog bred at Nate's Pond every year of our study. The first year we caught her she was huge, 105 mm and 154 g, suggesting she was old then. If, at that time, she was a 2-year-old first-time breeder, she would have been 10 years old at the end of our study. From her size when we initially caught her, however, it was likely she was at least 3 years old, and more likely 4 years or older when we first saw her. (We often discussed the question of Crawfish Frog size at specific ages, and decided for our population frogs 60–70 mm SVL were likely 1-year-olds, frogs 80–95 mm were likely 2-year-olds, and frogs longer than 100 mm were at least 3-year-olds, with females tending to be larger than males.) Before each field season we would ask each other if we thought we'd see Corner Burn Chick again, and it was always a celebration when we did. MJL learned to recognize her big, scarred body, but this was no trick—no other Crawfish Frog in our population looked like her (Figure 17.6).

17.4 FROG 780

Frog 780's primary burrow was located almost 1.2 km due west from Nate's Pond, where he had bred in 2009 (unlike females, who generally entered breeding wetlands gravid and exited spent, we could never determine whether males had actually bred or not). On 22 March, 2010, sometime after 9:30 PM, when Jen and I had checked his burrow and he was still there, 780 moved about 400 m straight east and uphill to a new wetland constructed to reduce erosion, and

FIGURE 17.5 Cohabitation of Frog 460's burrow by (from top to bottom) Frog 460 and a Gartersnake (black arrow), a small crayfish (white arrow), and a small rodent, likely a Deer Mouse. This burrow had two openings, and we called it the party burrow.

A **Daytime Activity**

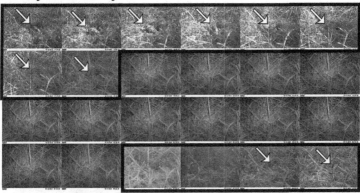

B **Nighttime Activity**

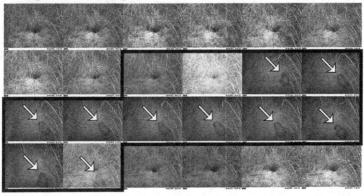

FIGURE 17.6 Frog 53 demonstrating diurnal activity on 12–13 October, 2012, and nocturnal activity two months earlier, on 4–5 August. Arrows indicate when she was outside her burrow.

stayed to breed. Jen estimated his pre-breeding migration speed to be about 33 m (100 ft)/hour. Eleven days later, on 2 April, Jen found his carcass on the bank, surrounded by raccoon paw prints in the mud.

17.5 FROG 060

In early 2009, MJL had the regrettable thought that rather than put radios in males leaving breeding wetlands, why not put them in males entering wetlands, in order to follow them while they were in the breeding wetland. Questions we could address with these data included the types of retreat sites males used during the day (in the water or on shore) and how often they switched locations during breeding choruses at night. One thing he did not think to consider was what effect the ability to produce vocalizations of over 100 dB that could carry over a mile under favorable conditions would have on the integrity of abdominal stitches. We could find no mention of this risk in the scientific literature, and so we implanted radios in five males entering breeding wetlands. It took a few days for Jen to realize that these animals had a bloated-looking appearance, and on 15 April we began retrieving them. Once we opened them up, we saw that the high abdominal pressure needed to produce loud calls had either broken the sutures or torn the tissue around the sutures, causing internal organs to herniate through the abdominal muscle wall. MJL operated on all five animals to replace the organs and re-suture the incisions; four of these surgeries were successful. He removed the radios from one of these animals but kept radios in the other three to determine their fate. Five months later, 150 days post-surgery, two of these animals were still alive. Frog 060 was one of these animals. We monitored him throughout all of 2009, and saw him as he entered and exited Nate's Pond to breed in 2010 and 2011, meaning he survived two years following his herniation surgery. After the 2011 breeding season, we never saw him again.

17.6 FROG 139

This frog had his burrow near a road and we used this convenience to our advantage as we frequently filmed him. In 2009, he bred in Big Pond; when he left we implanted a transmitter and on 19 May set up a wildlife camera to monitor his behaviors. Later that summer, he injured his left vocal sac and it wouldn't inflate when he upland called. Along with Romeo, Frog 139 switched burrows following the heavy rains of 10 October. The burrow he adopted, about 10 m due east of his primary burrow, seemed barely adequate to us (at the time MJL wrote in his field notes: "139 moved into a small, un-froglike burrow"), but he survived the winter in it. The next spring, 139 was one of the first frogs out of his burrow. Ten days later, on 1 March, he moved back to his primary burrow before heading to Big Pond to breed on 21 March. We followed him for all of 2010. In 2011, he was in Big Pond for a long time, almost a month. When he returned, his primary burrow had been occupied and capped by a crayfish. For about a week 139 wandered nearby, west of his old burrow, then disappeared.

17.7 4B08406161—SON OF ROMEO

On 8 June, 2012, two of our technicians, Tenia Wheat and Pete Lannoo, caught an 85 millimeter SVL male Crawfish Frog entering Cattail Pond at Bucket 27. The frog had a 2011 Cattail Pond toe clip. Tenia and Pete had been checking for metamorphosing juveniles so did not have a pit tag reader. Three days later, MJL caught this frog exiting Cattail and inserted pit tag 4B08406161 into him. During the process the frog urinated, and MJL placed him outside the fence, according to our protocol for frogs exiting wetlands. In his notes, MJL wrote that the frog looked pink and thin. Two days later, this frog showed up at Cattail again, trying to enter at the same bucket, 27, but this time he was dehydrated and had been severely injured. His upper and lower jaws were broken and his tongue lacerated, all along the same plane, and he could not retract the nictitating membrane of his right eye (Figure 17.7). Although many

FIGURE 17.7 Consequences of choosing the wrong burrow. This one-and-a-half-year-old frog (it had a 2011 Cattail Pond toe clip) entered Cattail Pond in early June 2012, during a drought, and appeared to be water stressed. Three days later it left the pond and MJL gave it a pit tag. Two days later, it again showed up at Cattail Pond, this time severely injured, as shown. Its upper and lower jaws were broken and his tongue lacerated, all along the same plane, and it could not retract the nictitating membrane of its right eye. We felt at the time that this frog, being water stressed and seeking relief, may have entered an occupied crayfish burrow and gotten lacerated by a crayfish pincher. Its injuries were so severe that MJL euthanized it.

scenarios for these injuries were possible (a raptor talon, for example), we felt at the time that this frog, being water stressed, may have entered an occupied crayfish burrow and gotten lacerated by a crayfish pincher (cheliped). His injuries were too severe to surgically repair, so MJL brought him back to the lab and euthanized him.

17.8 FROG 26—ROMEO

We first saw Romeo on 3 April, 2009, at 3:31 AM, entering Cattail Pond in Bucket 14, on a night when we captured 27 other Crawfish Frogs entering wetlands. Even then he was special, a big male measuring 112 mm SVL and weighing 148 g. He was the 79th Crawfish Frog we'd captured, and one of the biggest. Vanessa's crew gave him pit tag number 4B08253605 and released him into the wetland. Sixteen days later, he exited Bucket 1 at 10:29 PM weighing 135 g; he'd lost 13 g while in his breeding wetland. Jen brought him into the lab and implanted a radio with the frequency 150.279. He was her 26th telemetered frog. Depending on the context, back then we referred to him as either Frog 26 or 279. Later in 2009, we replaced his old transmitter with 149.129, and on 10 April, 2010, replaced that transmitter with 150.220. Twice in 2010, once on 28 May and again from 9–11 June, this male shared a single-opening burrow with Frog 33, pit tag number 4B07766F5C, a big female carrying transmitter 149.069. Following these cohabitations, Jen dubbed Frog 26 Romeo, and Frog 33 Juliet. In total, Jen tracked Romeo for 606 days. Perhaps no other animal has given up so many of its species secrets, while at the same time playing ambassador.

Romeo initially distinguished himself by switching burrows. While most other Crawfish Frogs Jen tracked from their breeding wetlands to upland burrows found a burrow and settled in, Romeo used burrows in a small area, and for a while after returning from breeding moved among them.

Romeo's tête-à-tête with Juliet provided the exception that proved a rule. Despite Reeve Bailey's assertion that Crawfish Frogs share burrows, in Jen's radiotelemetry study adult Crawfish Frogs cohabited single-opening burrows for only 8 of the 7,898 "frog days" she tracked radiotelemetered frogs (a cohabitation rate of one night every three years). During each of these eight nights, Romeo was with Juliet.

Romeo also hinted at why Crawfish Frog adults do not share burrows. On 9 June, 2010, Jen's notes read: "Romeo shacked up with Juliet." Two days later she wrote, "Romeo at Juliet's burrow. When I walked up either Romeo or Juliet jumped into the burrow but could only get in a few inches. I'm assuming the other frog is blocking the passage down." This is why Crawfish Frog adults inhabit burrows singly; if they spent much time cohabiting burrows their burrow-based defensive strategy would not work. Bailey's comment that Crawfish Frogs share burrows was based on observations of juveniles, and we have observed this, too. But juveniles occupy burrows with bores much larger than they are, and unlike adults, can move past each other when occupying burrows. From 27 March–15 April, 2012, our wildlife cameras captured images of a large adult Crawfish Frog sharing a double-opening crayfish burrow with a smaller adult. The large adult was established when the small adult appeared and remained after the smaller frog left. Each frog used a separate burrow entrance, and we observed no signs of aggression by either frog. Burrows with double entrances permit two Crawfish Frogs to each exhibit defensive behaviors in the presence of predators.

In 2010, Romeo alerted us to an unexpected phenomenon. While at our study site collecting data for other projects, we would set up the lab's digital camera to video frogs at their burrows. Because Romeo's burrow was conveniently located near where we parked to sample Cattail Pond, we spent a lot of time filming him. And in early September 2010, we happened to film Romeo calling in response to an airplane flying overhead.* This observation started us thinking about non-breeding vocalizations, and spun us off in a new direction, on an upland calling project which Nate led (Engbrecht et al. 2015). Three years later, Romeo filled in an important gap between upland and breeding calls, by showing this transition can be abrupt. In early 2013, we set up both a wildlife camera and a Song Meter audio recording unit on Romeo's burrow. On 11 March, 2013, images from the camera showed that Romeo emerged from his burrow at 11:54 PM and about an hour later began calling. He completed three calling bouts, beginning at 1:49 AM and ending at 2:40 AM. Calls from the first bout were upland calls; calls from the second bout alternated elements of upland and breeding calls; while calls from the third bout began as intermediate between upland and breeding calls, then graded into breeding calls. Immediately after producing these breeding calls, Romeo took off and began his migration to Cattail Pond.

The biggest insight Romeo gave to us came on 4 July, 2011 (Engbrecht et al. 2012) (Chapter 4). We had an old-style wildlife camera set up to record his activity hourly. At 2:00 PM, Romeo was on his feeding platform facing his burrow opening, as Crawfish Frogs almost always are. An hour later Romeo was absent—in his burrow—and a Black Racer had his head in the burrow, and appeared to be using his body to get leverage, really working to get at Romeo. The next four hours of images showed no snake, and no Romeo. At this point, there were only two possible outcomes and, being good field biologists, we

---

* See http://www.lannoolab.com/crawfish_frogs.htm#Crawfish_Frog_video.

were beginning to favor the worst-case scenario. Then, at 8:00 PM, Romeo re-appeared in exactly the same spot on his feeding platform facing exactly the same way as he had been at 2:00 PM—he was not bleeding and had no scars—looking unfazed. We talked about this photo sequence a great deal, and as we did, we came to realize how important burrow habitation is to Crawfish Frog predator avoidance (no other North American frog species could have survived such a direct attack by a snake). Jen's attempts to extricate Crawfish Frogs wedged in their burrows using the flushing technique she developed, and the Hognose Snake predation event filled in details about how these frogs avoid snake predation and the consequences of making a mistake. But the initial insight began on Independence Day, 2011, with a frog called Romeo.

One last story. Romeo became an ambassador for his species. On 29 February, 2012, after it warmed up but before frogs began their breeding migrations, MJL hosted Peabody Coal biologist Bryce West and the Indiana DNR's Division of Reclamation's Todd Sellers. They toured our study site and MJL pointed out the advantages of those early post-federal surface mining regulation restorations for Crawfish Frogs. They asked if they could see a Crawfish Frog burrow, so MJL took them to Romeo's burrow, and as luck would have it, he was out on his feeding platform. MJL slowly placed his left hand over Romeo's burrow entrance—his only escape route—and gently grabbed him with his right hand. Romeo offered no resistance, and was admired as he was passed around, hand to hand. MJL placed Romeo back on his feeding platform, facing his burrow entrance, where he sat as the group left.

In 2014, Romeo showed up at Cattail Pond and was lethargic. His belly and chin skin were red and scaly. We put him in Cattail, where he stayed a few days but didn't move beyond the area we'd placed him. On the chance that his skin lesions were a chytrid infection, on 7 April, we brought him into the lab and placed him in our incubator, which was set at 30°C. He succumbed, and we removed him and began palpating his heart in an attempt to revive this old frog who had taught us so much about his species. We got his heart beating again but could never get him to breathe on his own, and after about 45 minutes gave up and let him die in peace. As mentioned in Chapter 8, we texted Jen to let her know.

Is There Hope for the Hoosier Frog?

18

With all of this interesting biology out there just waiting to be discovered, it was easy to lose track of why we were doing this project in the first place—to assess the status of Hoosier Frogs in Indiana and make management recommendations to ensure populations survive in perpetuity. But these subjects were never far from our minds; Nate was in charge of this project, and we discussed this issue on an almost daily basis.

Early reports of Crawfish Frogs in Indiana date to the latter half of the nineteenth century. As mentioned in Chapter 2, F. L. Rice and N. S. Davis first reported Crawfish Frogs in Indiana in 1878 from a specimen collected by E. F. Shipman in Benton County (Chicago Academy of Sciences collection, CS 160) (Rice and Davis 1878). Willis Blatchley reported two additional specimens collected by C. Stewart and H. McIlroy from Vigo County in 1893 and 1894, respectively (Blatchley 1900).

R. Mumford, A. P. Blair, H. P. Wright and G. S. Myers of Indiana University (Wright and Myers 1927), and David and Paul Swanson, foresters for the Emergency Conservation Works and the Resettlement Administration, also contributed early Crawfish Frog records in Indiana (Swanson 1939). After he returned from the Second World War, Sherman Minton secured a number of new specimens, documenting the species presence in at least seven additional counties. In 1965, David Rubin reported Crawfish Frogs from a site now known as "Dave's Pond" in northern Vigo County (Rubin 1965). Over the years, numerous specimens collected from this site have been deposited in the Indiana State University Vertebrate Collection.

Sherman Minton provided the best early assessment of the distribution of Crawfish Frogs in Indiana (Minton 1972, 2001). According to Minton, Crawfish Frogs were considered "locally plentiful" in western Indiana until about 1970, when populations began to experience unexplained declines. He noted their disappearance at many localities, including sites with habitats that appeared to have experienced little change. By 1990, Mike Lodato had witnessed the extirpation of Crawfish Frogs from three sites near Evansville (Lodato, pers. comm.). One of these, located at Angel Mounds State Historic Site, apparently supported a population of >100 breeding adults before its numbers plummeted. Due to their increasing rarity in Indiana, Crawfish Frogs were designated a Species of Special Concern in 1984 and elevated to State Endangered status in 1988.

In March 2003, a group including Drs. Daryl Karns, Joseph Robb, and Hanover University student Erin Haswell confirmed the presence of a large population of Crawfish Frogs at Big Oaks National Wildlife Refuge (Haswell 2004). This discovery was a great surprise, added Jefferson, Jennings, and Ripley counties to the Indiana distribution, and extended the known range of Crawfish Frogs approximately 90 kilometers eastward. The source of this apparently isolated population is not known, and its status as a natural or introduced population has not yet been determined.

We wished to know the historic status of Crawfish Frogs in Indiana in order to provide a baseline to compare with current populations. To do this, Nate compiled a complete list of all known Crawfish Frog sightings in the state by examining all historic records and available specimens of Crawfish Frogs collected in Indiana. He used locality data deposited along with museum and university specimens, as well as literature accounts, Indiana Department of Natural Resources (IDNR) Division of Fish and Wildlife records, the IDNR Division of Nature Preserves Heritage Database Center, and other reliable reports.* Many of the most recent records came from an IDNR Crawfish Frog survey conducted from 2004–2008, which Nate had participated in as an IDNR technician (Walker, pers. comm.).

Nate then went into the field to check the status of populations indicated by the locality data he had compiled. He used a number of survey techniques including breeding call surveys (conducted both manually and using Song Meter audio recording units), egg mass surveys, incidental road kills, drift-fence surveys, aquatic surveys for tadpoles, and terrestrial surveys for newly metamorphosed juveniles. Most of Nate's data came from call surveys, which is not surprising. Crawfish Frog populations are most easily detected by the

---

* Engbrecht and Lannoo (2010). Nate obtained Crawfish Frog records from the following sources: Wildlife Diversity Section, Division of Fish and Wildlife, IDNR (IDNR WDS), Indiana Natural Heritage Data Center, Division of Nature Preserves, Indiana Department of Natural Resources, (INHDC), Indiana State University (ISU), University of Michigan Museum of Zoology (UMMZ), Field Museum of Natural History (FMNH), Chicago Academy of Sciences (CA), Harvard University Museum of Comparative Zoology (HUMCZ), Carnegie Museum of Natural History (CMNH), Texas Cooperative Wildlife Collection (TCWC), California Academy of Sciences (CAS-SU), and the Indiana State Museum (INSM). Data were obtained from records held in the following institutions and accessed through HerpNET data portal (http://www.herpnet.org): TCWC, 16 September, 2009; CAS-SU, 16 September, 2009; and CMNH, 20 August, 2009.

loud, distinct calls of breeding males (Swanson 1939; Gerhardt 1975; Williams et al. 2013), and thus offer the best opportunity for detecting populations. Nate surveyed from 2009–2011, including nine historic localities (Engbrecht and Lannoo 2010; Engbrecht et al. 2013).

To provide a rough assessment of population size at his surveyed wetlands, Nate estimated numbers of males present in wetlands based on maximum calling rate (Engbrecht and Lannoo 2010). At Hillenbrand Fish and Wildlife Area-West (HFWA-W), he estimated the overall population size using Vanessa's drift-fence and funnel-trap data. He and the biologists at Big Oaks NWR estimated population sizes using egg mass surveys at wetlands where calling males had been heard, which they assumed equaled the number of females. They then applied a 1:1 sex ratio to estimate population size. In cases where distant chorusing was heard at restricted sites at Big Oaks and egg masses could not be counted, Nate used chorusing levels to estimate population sizes.

Minton included 23 counties in the range of Crawfish Frogs in Indiana (Minton 2001). Sixteen of these counties were represented by point localities signifying reliable records and included Benton, Clay, Daviess, Dubois, Fountain, Greene, Martin, Monroe, Morgan, Owen, Pike, Sullivan, Vanderburgh, Vermillion, Vigo, and Warrick. Seven counties not represented by point localities included Warren, Parke, Putnam, Knox, Gibson, Posey, and Spencer. While not exhaustive, Minton's assessment provided the most thorough early compilation of Indiana Crawfish Frog distributional records. The IDNR audio surveys performed during the springs of 2004–2008 covered 17 counties, including the counties indicated by Minton plus Knox, Lawrence, Orange, and Park Counties. Crawfish Frogs were identified in seven: Vigo, Clay, Owen, Daviess, Sullivan, Parke, and Greene. The IDNR surveys failed to detect Crawfish Frogs at sites in Morgan, Monroe, Fountain, Vermillion, Pike, Martin, and Dubois Counties, where Minton had indicated extant Crawfish Frog populations.

Nate's surveys were more extensive than the IDNR audio surveys, relying on a variety of sampling techniques. Nate summarized his conclusions in the context of historic records, by county, as follows (Figure 18.1).

Benton County—Crawfish Frogs were first reported in Indiana from Benton County in 1878. This record represents the type locality for the subspecies *Rana areolata circulosa*, the Northern Crawfish Frog, and the northernmost locality for Crawfish Frogs in the state. Subsequent to Nate's statewide assessment, Resetar and Resetar provided an in-depth examination of the provenance and collection date of Shipman's original Northern Crawfish Frog specimen. Based on biographical and botanical records, they concluded the original Benton County "Hoosier Frog" likely was collected during the summer or fall of 1876, possibly in Grant or Gilboa townships where the Shipman family farms were located (Resetar and Resetar 2015).

Presumed extirpated
(pre-2000 records only)

Potentially extirpated
(2000–2008 records only)

Extant
(post-2009 records)

FIGURE 18.1 Our current understanding of the distribution of Crawfish Frogs in Indiana. Shaded counties represent all documented populations. Black-shaded counties represent counties that currently support populations.

Clay County—Minton included two point localities for Crawfish Frogs in Clay County, and noted hearing a "spectacular chorus" on 2 April, 1950. A Clay County specimen Minton collected on 2 April, 1950 (UMMZ 101623) likely corresponded to the location of the large chorus he heard that night. Russell Mumford collected a Crawfish Frog from northern Clay County on 7 April, 1958 (UMMZ 118078). David Rubin and P. Allen collected a specimen near Bowling Green on 18 April, 1966 (ISU 1492). IDNR personnel reported hearing Crawfish Frogs near Brazil on 26 March, 2007. Nate surveyed for Crawfish Frogs near Brazil where IDNR personnel identified chorusing between 2004 and 2008, but did not detect them. He also conducted calling surveys in areas of extensive grassland habitat at Chinook Fish and Wildlife Area but, again, did not detect them.

Daviess County—Paul Swanson frequently heard Crawfish Frogs from "within the city limits of … Odon" (Swanson 1939). Minton and W. M. Overlease collected a Daviess County specimen on 21 March, 1953 (UMMZ 108125). IDNR biologists reported a cluster of four call points northeast of Odon between 2004 and 2008, which appeared to be distinct from the sites reported by Swanson, Minton, and Overlease. These surveys also revealed a cluster of breeding sites in south-central Daviess County.

Nate confirmed the presence of Crawfish Frogs in two distinct population clusters in Daviess County, one northeast of Odon, the second in the south-central region identified by the IDNR surveys. He did not detect Crawfish Frogs at four sites identified by IDNR between 2004 and 2008. Similarly, Nate heard no calling at the site east of Odon surveyed by Minton and Overlease in 1953.

Dubois County—Swanson identified Crawfish Frogs from Dubois County and characterized them as "quite plentiful" (Swanson 1939). IDNR surveys from 2004–2008 failed to detect this species, and therefore the status of Crawfish Frogs in Dubois County is uncertain.

Fountain County—Fountain County is represented by a single vouchered specimen (FMNH 64663) collected near Kingman by Minton on 18 April, 1951. This animal was reported from a shallow pond in a cultivated field (Resetar, unpubl. data). IDNR surveys from 2004 and 2008 failed to detect Crawfish Frogs at this site. Nate also surveyed this area and heard no calling.

Greene County—There were at least 17 Crawfish Frog reports from Greene County, most based on the recent IDNR surveys. Minton deposited an animal collected near Worthington on 25 March, 1949, in the University of Michigan Museum of Zoology (UMMZ 100304). Nate surveyed Minton's site, but could not detect Crawfish Frogs. Nate also did not detect Crawfish Frogs at four IDNR sites near the towns of Scotland, Linton, and Jasonville.

Nate identified several new Crawfish Frog localities in Greene County, most notably at HFWA-W, our study site. We have now detected chorusing at a total of nine wetlands at HFWA-W, which likely supports the densest assemblage of Crawfish Frogs in Indiana.

In 2002, Matt Blake and Vicky Meretsky first reported Crawfish Frogs from the Goose Pond basin, south of Linton. IDNR surveys from 2004–2008 identified Crawfish Frogs in six areas within the Goose Pond basin, including a confirmation of the Blake and Meretsky record. Nate's more recent surveys, however, suggest that Crawfish Frogs have precipitously declined at Goose Pond. He detected Crawfish Frogs at only two sites, both on adjacent private property. Nate did confirm light chorusing (one or two males) at a new site on Goose Pond FWA several kilometers from the two sites previously mentioned (Sternenburg, pers. comm.). The populations at these three Goose Pond sites are small and on private land, and appear to be at risk.

Martin County—Swanson included Martin County in a list of counties where Crawfish Frogs were "quite plentiful" (Swanson 1939), and reported frequently hearing them from within the city limits of Loogootee (Swanson 1939). He did not voucher any specimens. Nate visited Loogootee but could not detect the presence of Crawfish Frogs.

Monroe County—Wright and Myers reported finding a population "two miles west of Bloomington" on 21 March, 1926 (Wright and Myers 1927). They collected specimens and deposited them in the California Academy of Sciences (CAS-SU 2174–2180, 13343–13364). Mittleman reported H. T. Gier collected one juvenile and an unknown number of tadpoles from a small pond "four miles north of Bloomington" on 12 April, 1940 (Mittleman 1947). These specimens were deposited in the Ohio University collection (OUZ A1126), but appear to have been relocated and may now be lost. A. P. Blair deposited a series of metamorphosing tadpoles at the University of Michigan Museum of Zoology (UMMZ 95312) dated 19 July, 1940 with the locality description of "Bloomington." These animals may have been Gier's. Minton apparently considered the localities reported by Wright and Myers, and Mittleman to be the same population (Minton 1972). On 23 March, 1991, Al Parker reported sighting two Crawfish Frogs at a wetland near Bloomington along the Beanblossom Creek bottoms. During their 2004–2008 surveys, IDNR biologists were unable to confirm their presence, despite numerous visits, and Crawfish Frogs were presumed to be extirpated from this location. Nate conducted surveys at these historic sites and also did not detect Crawfish Frogs. While suitable open, grassy habitat was present, the large wetlands, particularly those that connect with Beanblossom Creek during high water, appeared to contain fish, which would make them unsuitable for Crawfish Frog reproduction.

Morgan County—In early April 1978, Robert Luker collected two Crawfish Frogs (INSM 71.7.170–171) from Monrovia. These records appear to represent the easternmost vouchered record in this species' contiguous range in Indiana. This population appears to have persisted until at least 1987 (Engbrecht and Lannoo 2010), but IDNR surveys from 2004–2008 failed to detect Crawfish Frogs. We know of no current populations remaining in Morgan County, where Crawfish Frogs may now be extirpated.

Owen County—Minton collected Crawfish Frogs from Owen County on 25 March, 1954 (UMMZ 110638). In 2011, Nate conducted surveys at this site and heard Crawfish Frogs calling from two different wetlands. The first consisted of a degraded cattle pond. The second, which he could not definitively locate, appeared to be situated about 1 kilometer away. Both sites were located in a relatively large series of pastures and grassy fields.

Parke County—IDNR surveys identified a single population of Crawfish Frogs in Parke County on 26 March, 2007. The locality description associated with this record is vague and the exact location of this site is unknown. Nate could not find it and considered the status of Crawfish Frogs in the county to be undetermined.

Pike County—Swanson and Swanson collected a series of Crawfish Frogs from Winslow that are now deposited in the Carnegie Museum of Natural History (CMNH 13371–13375) (Swanson 1939). On 25 June, 1963, John Tritt collected a single Crawfish Frog "near Spurgeon" (ISU 2473). IDNR surveys from 2004–2008 did not detect Crawfish Frogs in Pike County, and Nate considers their status here to be unknown.

Spencer County—D. S. Dougas discovered Crawfish Frogs near Newtonville in 1998 (Lodato, pers. comm.). Frogs at this site appear to be using a series of breeding ponds situated over several acres of reclaimed mine land. No voucher specimens have been collected, but biologists have visited this area—which consists of rolling grasslands and wetland swales, and is in private ownership—each year since, and have confirmed these populations persist (Lodato and Dugas 2013). In March 2008, Mike Lodato found evidence of a second population, near Chrisney, located about 6.5 kilometers from the first. Lodato found the animal on State Route 70 during a nighttime rainstorm, and photographed it. This photo is cataloged in the Illinois Natural History Survey collection (INHS 2011n), and serves as the voucher for Crawfish Frogs in Spencer County. More recently, in his capacity as Indiana state herpetologist, Nate worked with Mike Lodato to finally locate the source of this population.

Sullivan County—Sullivan County contains at least 26 Crawfish Frog records, with most occurring in the east-central region. Vouchered records include a specimen collected by Minton on 21 March, 1952 near Shelburn (UMMZ 105544) and a single adult collected by John Whitaker, Jr. near Sullivan during the first week of June 1969 (ISU 2255). In her report on the amphibian use of reclaimed and un-reclaimed surface coal mines, Anne Timm identified 14 Crawfish Frog localities representing a variety of habitats including a ditch, slough, beaver impoundment, and larger "final cut" strip pits (Timm 2001). Timm collected no voucher specimens and the current status of Crawfish Frogs at these sites is unknown.

IDNR surveys from 2004–2008 revealed nine Crawfish Frog localities from Sullivan County, including sites near Cass, Hymera, and Dugger. Most of these sites represent call points located along roadways. IDNR property manager Ron Ronk reported hearing Crawfish Frogs calling from a private wetland complex north of Dugger every year from 2004–2008 (Ronk, pers. comm.). Stuart Smith reported finding a Crawfish Frog after a hard rain near Lake Sullivan on 20 May, 2002. Voucher specimens are not available for these records.

Nate documented two new Sullivan County breeding sites in 2011, both near a large, reclaimed coal mine in the eastern portion of Sullivan County. The first site, called to Nate's attention by retired IDNR biologist Roger Stonebraker, is located in managed grassland on private property. This site was converted from agriculture to grassland in 2000, with wetland construction taking place around 2003 (Stonebraker, pers. comm.). Chorusing has occurred at this site every year since 2009, and choruses have intensified each year. The second site, approximately 4.5 km away, is in an agricultural field near Dugger. Frogs at both sites are likely using nearby state-owned grasslands as terrestrial habitat. Nate also re-confirmed two previously known populations in Sullivan County. The first came from a cluster of breeding sites near Hymera where we heard chorusing each year from 2009–2011. The second was reported by retired IDNR biologist Ron Ronk, and confirmed in 2011.

In addition, Nate surveyed for Crawfish Frogs at Minton's 1951 collection site near Shelburn, but did not detect them. This area is heavily farmed and little natural upland or wetland habitat remains. Nor did Nate hear Crawfish Frogs at ten sites where they had previously been reported in Sullivan County. Many of these sites were originally identified by Timm or by IDNR personnel at the now defunct Minnehaha FWA. Similarly, Nate's surveys here and at Greene-Sullivan State Forest's Dugger Unit failed to reveal any populations.

Vanderburgh County—P. L. Swanson and D. C. Swanson collected a Crawfish Frog on Route 41 in Vanderburgh County on 28 March, 1936, and deposited it in the Carnegie Museum of Natural History (CMNH 13378) (Swanson 1939). Three other Vanderburgh County sites are known to have supported Crawfish Frogs. Angel Mounds State Historic Site once supported a robust population containing an estimated 100 adults in 1980. By 1987 this population had shrunk to fewer than ten breeding individuals. By 1990 it was extirpated (Lodato, pers. comm.). Two nearby sites located in Evansville were destroyed by suburban development shortly after the extirpation of the Angel Mounds population. Nate suspects Crawfish Frogs have been extirpated from Vanderburgh County.

Vermillion County—Minton collected a specimen on 18 April, 1951 from a "shallow pond" near Perrysville in northern Vermillion County (UMMZ 103361), which represents the only known Crawfish Frog location for Vermillion County. Nate spent a great deal of time surveying this site, but failed to detect Crawfish Frogs. The historic Vermillion and Benton County records are the two northernmost records in Indiana, and the only Indiana populations known to occur west of the Wabash River.

Vigo County—Crawfish Frogs were first reported in Vigo County by Blatchley, who received two specimens collected by C. Stewart at "the south part of the city of Terre Haute" on 8 and 9 October, 1893. H. McIlroy collected a third specimen "three miles west from where the others were secured" on 9 May, 1894 (Blatchley 1900). Locality data for these sites are vague. The two Vigo County specimens obtained by Blatchley were deposited in the Harvard University Comparative Museum of Zoology (HUMCZ A-7043, A-7044) and have a collecting date of 9 October, 1903. Either these are the same specimens collected by C. Stewart (same date but the year mistakenly listed as 1903), or they are new specimens coincidentally collected on the same calendar date (9 October).

On 24 March, 1964, Dave Rubin discovered a new population of Crawfish Frogs in northeast Vigo County (Rubin 1965). This area, now called Dave's Pond, contains at least three separate wetlands and has been visited numerous times over the past several decades by researchers from Indiana State University. A number of voucher specimens have been collected from this site (ISU 395–397, 399–400, 401–403 [eggs only], 937, 2738, 2793, 2822, 3177 [eggs only], 3204–3207; PU 8482–8483). A specimen collected by E. G. Zimmerman on 6 April, 1964 (TCWC 66467) contains the locality description "5 mi NE Terre Haute" and may correspond to the Dave's Pond complex. Nate visited this site and detected Crawfish Frogs each year from 2009–2011. Dave's Pond currently represents the only known extant population in Vigo County, and the northernmost Crawfish Frog population in Indiana.

John Whitaker and Dave Rubin collected a specimen about 3 miles ENE of Dave's Pond near Fontanet on 30 March, 1967 (ISU 1820). In the late 1960's a frog was observed in the base of a broken metal pole about 3 miles west of Dave's Pond (Whitaker, pers. comm.). Nate did not detect Crawfish Frogs during surveys at this site, which is now a matrix of agriculture, woods, and grassland habitat.

IDNR biologists identified a Crawfish Frog breeding site near the Parke County line in 2007. This location is situated in a low, flat basin near Raccoon Creek. A specific location was not identified, but a series of small wetlands were present. Nate repeatedly visited this site but failed to detect Crawfish Frogs.

Warrick County—Swanson included Warrick County in a list of counties in which Crawfish Frogs are described as being "quite plentiful" (Swanson

1939). However, Minton and his colleagues noted that "some colonies in Vanderburgh and Warrick counties have been destroyed by surface mining, drainage, and urban expansion" (Minton et al. 1982). Lodato reported Crawfish Frogs from three sites near Elberfeld, Millersburg, and Paradise that were destroyed by mining operations (Lodato, pers. comm.). No Warrick County specimens have been vouchered, and no extant populations remain.

Jefferson, Jennings, and Ripley counties. Records for Jefferson, Jennings, and Ripley counties are all located within Big Oaks National Wildlife Refuge, and thus have been placed together here. The suspected presence of Crawfish Frogs at Big Oaks in the spring of 1999 was finally confirmed in March 2003 (Hauersperger 2005). Two specimens collected by Daryl Karns, Joseph Robb, Erin Haswell, and Diana Schuler on 18 March, 2003 were deposited in the Field Museum of Natural History (Jefferson Co: FMNH 262589; Ripley Co: FMNH 262588). Haswell conducted a survey and identified 23 sites at Big Oaks: 21 breeding call locations, two sight records (Haswell 2004). We do not know whether Crawfish Frogs naturally occurred here or whether personnel at the former Jefferson Proving Grounds introduced them (Haswell 2004). The genetics work we've done so far have proved inconclusive (Nunziata et al. 2013).

Joe Robb's team systematically searched portions of Jefferson, Jennings, and Ripley counties contained within Big Oaks NWR, and opportunistically searched areas outside the refuge using call surveys to locate Crawfish Frog breeding choruses. They confirmed breeding at 15 Jefferson County wetlands. Three of these sites were discovered in 2009, one in 2010, and nine in 2011. Calling at two wetlands had been heard in previous years (one wetland in 2004, one in 2007). They detected between one and 15 Crawfish Frog egg masses at these sites. Breeding ponds varied, ranging from small round bomb craters (Big Oaks NWF used to be the U.S. Army's Jefferson Proving Grounds) to large, shallow, flat-bottomed wetlands. All breeding ponds were in grassland habitat, with the exception of one pond located in a late successional deciduous forest (Williams et al. 2012).

Crawfish Frogs appear to have been locally extirpated at two sites in the Jefferson County portion of Big Oaks NWR. Robb's crew heard chorusing at the first site in 2008 and trapped a single male in 2009, but did not detect Crawfish Frogs in 2010 or 2011. They have not heard Crawfish Frogs calling from the second site since 2006. Within the Jennings County portion of Big Oaks, they confirmed four Crawfish Frog breeding wetlands. They first heard calling at two sites in 2008 and two others in 2011. Each wetland had between three and five egg masses. All four wetlands were small (<50 m^2), shallow (<1 meter), and situated in grassland habitat. They confirmed seven Crawfish Frog breeding wetlands in the Ripley County portion of Big Oaks, where they first heard calling at two wetlands in 2004, one in 2010, and the other four in 2011.

They also heard Crawfish Frog calling within the restricted area of the Indiana Air National Guard, Jefferson Range, which is surrounded by Big Oaks NWR. Several of these locations supported large choruses.

18.1 THE STATUS OF CRAWFISH FROGS IN INDIANA

Crawfish Frogs historically occurred across the southwestern quarter of Indiana, ranging from the Ohio River north to Fountain and Benton Counties. In 2003 they were discovered at Big Oaks National Wildlife Refuge in the southeastern portion of the state. Formerly described by Minton (2001) as being "locally plentiful," Crawfish Frog declines led to their listing as state endangered species.

Nate's survey and statewide review identified Crawfish Frogs at fewer than 60 sites throughout the state, including 21 localities in southwestern Indiana and 27 localities at Big Oaks NWR in southeastern Indiana. Nate detected Crawfish Frogs at only one of nine historic sites, and despite his best efforts could not detect them at 25 sites where they had been reported as recently as 2000–2008.

Nate's data suggest remaining Crawfish Frog populations in Indiana are concentrated in two areas, in Greene and Sullivan counties (especially HFWA-W) in the southwest, and at Big Oaks NWR in the southeast. Hillenbrand FWA-W populations are few (seven), but concentrations of breeding adults are high (>70 adults at Nate's Pond; >100 adults at Big Pond). In contrast, densities of breeding adults at Big Oaks NWR are low (generally <30 adults/wetland) but populations are numerous (>27). Outside of these two public areas, populations are small—from estimates of 4 in the smallest to 48 in the largest—scattered, and generally occur on private property. Our data suggest perhaps 1,000 adult Crawfish Frogs remain in Indiana. Big Oaks NWR and HFWA-W each support about 300 breeding animals, and fewer than 400 breeding animals persist in the remaining populations, mostly on private lands.

While Nate's sampling strategy involved visiting the majority of recent Crawfish Frog localities and many of the historic sites in Indiana, we recognize that undocumented populations may remain. However, even doubling our current estimated number of breeding adults in Indiana to 2,000 places the estimate at only a fraction of the number of eggs contained in a single Crawfish Frog egg mass (~7,000, see above), which is a frightening statistic.

A few details about the contraction of the range of Crawfish Frogs in Indiana. There are no data to suggest the few northern populations in Benton, Fountain, and Vermillion counties continue to persist. In the south, most known populations from Indiana's Ohio River border counties (Vanderburgh, Warrick, and Spencer) have been extirpated; Crawfish Frogs here now occur in one or two small populations in Spencer County. Crawfish Frog populations in Morgan and Monroe counties may also be extirpated. Collectively, these declines have produced a range contraction in the northern, southern, and eastern portions of the historic range of Crawfish Frogs in Indiana. Our data indicate that the status of Crawfish Frogs as State Endangered in Indiana is currently warranted. Crawfish

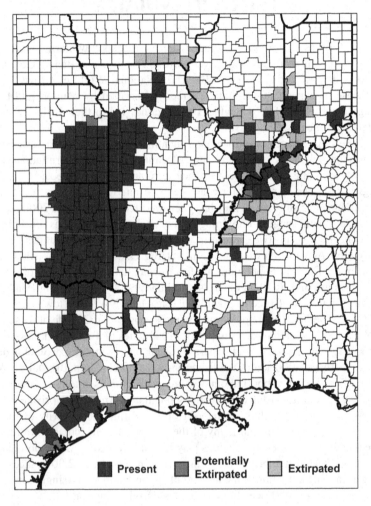

FIGURE 18.2 Our current understanding of the distribution of Crawfish Frogs throughout their range. Shaded counties represent all documented populations. Black-shaded counties represent counties that currently support populations.

Frogs appear to have been extirpated in 11 of 20 counties where they historically occurred. These losses have been partially offset by recent recolonizations of large restored grasslands at HFWA-W and Big Oaks National Wildlife Refuge.

Crawfish Frogs are a species of concern in each of the 13 states that host, or have hosted, populations. Nate, wishing to put his Indiana data into context, then asked if the shrinking and shredding of the Crawfish Frog distribution in Indiana he documented was being repeated in states across its range. We unearthed the digital Crawfish Frog county-level distribution map MJL had prepared for the 2005 *Amphibian Declines* book (Lannoo 2005) and we starting emailing it and passing it around at scientific meetings, especially the Southeastern PARC (Partners in Amphibian and Reptile Conservation) and Kansas Herpetological Society annual meetings. After Nate graduated, Jonathan Swan continued to distribute Nate's map and request comments and corrections, and today we feel pretty good about the overall picture this map presents (Figure 18.2).

We have evidence Crawfish Frogs occurred in 243 counties across the 13 states encompassing their range. Local expert opinion suggests they have been extirpated from 84, a 35% drop in county-level occupancy. If we divide their distribution into counties east and west of the Mississippi River, east of the Mississippi, Crawfish Frogs have been extirpated from 52 of 86 counties, a 59% loss in county-level occupancy, west of the Mississippi, Crawfish Frogs have been extirpated from 32 of 157, a 20% loss. From these data, and simply eye-balling the map, it's clear that the majority of Crawfish Frog declines have occurred east of the Mississippi River, in a pattern resembling what Nate found for Indiana. Crawfish Frogs are now categorized as "near threatened" globally, where populations are declining out of proportion to syntopic wetland breeding amphibians.

18.2 HABITAT REQUIREMENTS

The pattern of extirpation and colonization of Crawfish Frogs in Indiana is easily understood from their biology. Crawfish Frog populations require three landscape features:

1) Expansive grassy terrestrial habitat. Crawfish Frogs spend most of the year occupying burrows in open, grassy habitat, and will use the same burrows year after. Because of this fidelity to specific burrows, conserving terrestrial habitat is critical. We recommend landowners and land managers restore and/or maintain open grassy habitat when possible. In order to preserve the structural integrity of

burrows, it is critical that land managers on sites that host Crawfish Frogs avoid plowing and disking. If plowing is required for installing food plots or firebreaks, we recommend that strips be located as far from wetlands as possible, and that the same areas be plowed year after year.

In Indiana, extant Crawfish Frog populations are associated with several forms of open habitat including managed prairie, grassy meadow, abandoned field, hayfield, shrub land, and livestock pasture. Prescribed burning is frequently used in grassland management plans to control woody vegetation. Despite the aggressive spring and fall burning program at HFWA-W, including a 2017 arson event that torched most of the property, we've never observed a Crawfish Frog killed by a prescribed burn, although no doubt this can and does occur, especially during breeding migrations.

2) Fishless, seasonal or semi-permanent wetlands. Crawfish Frogs depend on fishless bodies of water for breeding. If Crawfish Frog populations are to persist, fish introductions must be avoided. To prevent natural colonizations of predatory fishes, wetlands designed to augment Crawfish Frog populations should not be constructed in areas subjected to riparian flooding. Seasonal drying of ephemeral wetlands will naturally eliminate established fish populations.

3) The presence of burrowing crayfish. Crawfish Frogs have a close association with burrowing crayfish and depend on them for the construction of their subterranean burrows. Thoma and Armitage note that five species of primary burrowing crayfish occur in Indiana, two of which, the painted-hand mudbug (*Cambarus polychromatus*) and the devil crayfish (*C. diogenes*), occur statewide (Thoma and Armitage 2008). Delineating management guidelines for burrowing crayfish is beyond the scope of this chapter, however Thoma and Armitage note that no burrowing crayfish species is currently of conservation concern in Indiana.

18.3 MANAGEMENT RECOMMENDATIONS

The biggest threat to Crawfish Frog populations has been habitat loss, and this threat may be intensifying as remaining populations on private lands become smaller and more isolated. Indeed, the pattern of decline of Crawfish Frog populations in Indiana suggests that in perhaps the next half-century most, if not all, remaining populations will be on public lands managed by state or federal biologists. It is essential that Crawfish Frogs become a component of the management plans in these areas if they are to avoid being extirpated in the state.

While in Indiana, the relatively large number of frogs on a handful of public lands provides a buffer against threats such as disease; the smaller, scattered populations on private lands may function to preserve genetic diversity. State and federal biologists who regularly work with private landowners, including coal companies, can help secure remaining populations by providing assistance to preserve these small, scattered populations.

Finally, because of the recent work detailed above, the prospect of restoring populations in Indiana is now within reach. Since Crawfish Frogs will colonize new sites where adequate habitat is available (whether through natural dispersal or human-assisted reintroductions), land managers have the opportunity to expand and establish populations by managing for grassland ecosystems. If Crawfish Frogs are incorporated into both public and private land management plans, the grasslands in the southern portion of the state are extensive enough to at least double the number of populations, which would enable state biologists to downlist this species from Endangered to Special Concern.

The Answers Are Blowin' in the Wind

19

There remain at least five big gaps in our knowledge of Crawfish Frogs. The first is the question of subspecies, which is actually two questions: are the existing subspecies designations real, and if so where are their geographic boundaries? Arthur Bragg (1953) reported clear subspecies boundaries in Oklahoma, where Southern Crawfish Frogs occur in the southeastern corner of the state, and Northern Crawfish Frogs occur in the east-central and northeastern portions of the state. Dowling (1957), and later Stan Trauth and his colleagues (Trauth et al. 2004) also reported clear subspecies boundaries in Arkansas, where Southern Crawfish Frog populations occur only in the extreme southwestern counties (Miller and Lafayette), while northern Crawfish Frog populations are scattered along the western half of the Arkansas River tapering off east, with other clusters of populations occurring in the northwest and much less densely in the southeast. Southern Crawfish Frogs appear to be the only subspecies in Texas and Louisiana, while Northern Crawfish Frogs appear to be the only subspecies in Kansas, Missouri, Iowa, Illinois, Indiana, Kentucky, Tennessee, and Alabama.* The problem with this scenario is that the distribution of Crawfish Frogs appears as a Japanese fan anchored in eastern Texas, where the proximal portion of the fan represents Southern Crawfish Frogs, the distal portions Northern Crawfish Frogs. The impression is that Southern Crawfish Frogs in southern Oklahoma, for example, evolved into Northern Crawfish Frogs in northern Oklahoma, and Southern Crawfish Frogs in eastern Louisiana evolved into Northern Crawfish Frogs in Alabama. But evolution doesn't work this way. All Northern Crawfish Frogs must be more closely related to each other than they are to any Southern Crawfish Frogs. If this is not true, the subspecies designations as currently depicted are invalid.

There are clear geographic differences in size across the range of Crawfish Frogs, with largest animals reported in the north, and the smallest in

---

* For a map, see Powell et al. (2016).

the south. (MJL caught a breeding adult Southern Crawfish Frog at Attwater Prairie Chicken National Wildlife Refuge in Texas that was 72 mm SVL. Had he caught this animal in Indiana he would have considered it a juvenile. Not everything is bigger in Texas.) Are these true subspecies effects, or is this size difference a function of ecological factors related to latitude?

A second gap in our knowledge revolves around the difference in triggers to breeding across Crawfish Frog populations. While Crawfish Frogs in the north are highly seasonal breeders, Crawfish Frogs inhabiting the Gulf Coastal Plain appear able to breed anytime from October through March following heavy rains. In fact, the breeding pattern of Coastal Plain Crawfish Frogs more closely resembles the breeding patterns of Gopher Frogs than it does the breeding pattern of inland Crawfish Frogs. It would be interesting if these differences in triggers to breeding broke out along subspecies lines confirmed by genetic analysis. If so, there would be ample justification for splitting Crawfish Frogs into two species.

A third gap in our knowledge involves the nature of the Crawfish Frog breeding system. In ranid frogs, males find suitable egg deposition sites and begin calling to attract females. Bullfrog males will establish physical territories, much like songbirds do, but most other North American ranid males will select a site, call, and defend this area. Large males fight and dominate smaller males, whose only reasonable chance of mating is to remain silent and ambush females on their way to mate with calling, dominant males—a strategy termed "sneaky fuckers" by Dawkins and Krebs in Krebs and Davies influential 1978 book *Behavioural Ecology*.* We believe Crawfish Frog mating behavior follows this lek-like pattern, but our evidence is circumstantial, based on egg deposition sites, calling behavior, and finding injured males, as follows.

As Bragg observed, Crawfish Frog egg deposition sites tend to be clustered around calling sites. We can add that calling and egg deposition sites vary from year to year. These observations suggest the presence of some strong attractor, such as a large calling male, that shifts annually. Furthermore, as male numbers in breeding wetlands increase, calling activity rates do not keep pace (Williams et al. 2013), suggesting that not all males present are calling. Finally, we occasionally observed males whose vocal sacs were ruptured and bloody, suggesting male–male aggression. Gloyd observed:

> Occasionally one [male] would make a rush at another which would evade the plunge by deflating itself and making a sudden dive beneath the surface of the shallow water, immediately coming up only a few inches away and again participating in the same behavior. These actions although indulged

---

* Dawkins and Krebs (1978). And while it is true that their description is inaccurate for most frogs—the only frogs that have intercourse are Tailed Frogs—we are determined to resist any and all politically correct attempts to clean up biology.

in by males only did not seem to have the nature of combat, but rather of a friendly game of sport. At these times females apparently in the role of spectators [ugh] were seen at the edge of the water or higher on the bank of the pool.

While none of these behaviors by themselves or collectively proves a lek-like reproductive system, they are suggestive. If genetic studies comparing the genomes of males to egg masses showed that the largest males fertilized a disproportionate number of egg masses, the evidence for lek-like behavior would be stronger than any field data we've collected to date.

A fourth gap in our knowledge is when, during their juvenile stage, Crawfish Frogs eventually settle into a primary burrow—one that fits the girth of their body and can be used to assist in predator defense. We know from our telemetry studies that burrow adoption does not happen immediately following metamorphosis—that the post-metamorphic urge is to move and disperse, not find a retreat site and settle in. We also know that even when juveniles occupy burrows, they range many body lengths away from the burrow entrance, much as Archie Carr has described for adult Gopher Frogs.* Further, we know from our wildlife camera surveillances that one-year-old juveniles also do not adopt burrows that fit their girth, instead they occupy large burrows on a temporary basis. They will also share burrows with other juvenile Crawfish Frogs more often than we've observed burrow sharing in adults. Although we have no direct evidence for this, our sense is that by the end of their second summer (i.e., when they are a year-and-a-half old), Crawfish Frog juveniles begin adopting primary burrows. This is also the time when the females who will be breeding in their second year begin developing eggs.

The final large gap in our knowledge involves the cues Crawfish Frogs use to orient themselves across their landscape. Animal orientation has always been a difficult question to address, let alone answer. For example, in *Behavioural Ecology*, the first book to address this discipline (Krebs and Davies 1978), homing is not mentioned at all, and dispersion and migration is given a page and a half (out of almost 500 pages of text). And while many of today's scientific papers consider magnetic fields to be important to amphibian orientation across the landscape (for example Phillips et al. 2002; Diego-Rasilla et al. 2008), the neuroscience-based authority (Butler and Hodos 1996, p. 35) states "Although the evidence for magnetoreception is convincing, where these receptors are and what they might be remains a mystery." So, we know that amphibians can find their way home (Crawfish Frogs being one of the best

---

* Carr (1940, p. 64): "[Gopher Frogs] range some distance from their retreats in foraging at night … I once saw one seated on a fallen pine log; looking about I located a gopher-hole some thirty feet from the frog; when I advanced and kicked one end of the log the frog hopped about wildly for a moment, then headed directly for the gopher hole and disappeared down it."

examples), we just don't know which sensory systems they use to accomplish this. Further, across species, especially between frogs and salamanders, different cues are likely to be used. (For example, while many species of troglodytic blind salamander species have evolved, despite there being 12 times more frog species than salamander species, there are no troglodytic blind frogs, suggesting that vision is a much more critical sense in frogs than it is in salamanders.)

Much of our knowledge on amphibian orientation and homing comes from observational and experimental manipulations—for example displacing animals to some novel site and observing their movements, or bringing animals into the laboratory and exposing them to altered stimuli such as modified magnetic cues. Because of their status as state endangered in Indiana, we couldn't do these sorts of manipulations on our Crawfish Frogs. Further, given their dependence on burrows for defense, it's likely that displacement experiments will expose Crawfish Frogs to predators, which will lead to their death. (If such studies are done, we recommend using young males, since they seem to be the most expendable.)

What we can say at this point is that because Crawfish Frog breeding adults migrate most often at night, during rains, any cues they use to find their primary burrows must be present at night, or if present during the day must be remembered and then applied at night. We also know from Jen's telemetry work that Crawfish Frogs can have a difficult time locating their burrows once they get close to them (compare Figure 4.3; post-breeding (left and right) panels to the pre-breeding (center) panel). This may mean that they use different cues (something like magnetoreception) to get into the vicinity of their burrows, then switch to visual landscape cues to home in.

Our data also suggest that Crawfish Frogs do not consider the migration as a whole, but rather split it into segments. Jen found that 80% of the retreats Crawfish Frogs used during their post-breeding migration were ones they had used previously during their pre-breeding migration. It is not difficult to imagine Crawfish Frogs remembering the route used during their pre-breeding migration, and following it backward, including using retreat sites, during the post-breeding migration.

While some amphibians will only orient themselves within an environment they know (Sinsch and Kirst 2016), others can orient from much farther distances (Pasukonis et al. 2018). In this context, we wish we knew more about juvenile movements, especially during their second year. It is possible that juveniles spend their first year dispersing (see Chapter 13), and their second year searching for a suitable burrow. These juvenile movements in combination with the ranging movements Jen documented, may provide adult Crawfish Frogs all the information they need to know about their landscape and their place in it.

As you can see, there remains much work to do before we can truly say we understand the biology of Crawfish Frogs. But, happily, data are being gathered from other populations across the range of this species. For her master's research at Stephen F. Austin State University, Laura Springer radiotracked 24 Southern Crawfish Frogs at Attwater Prairie Chicken National Wildlife Refuge (Springer 2016; Kwiatkowski et al. 2016). Similar to our results, she found that Southern Crawfish Frogs used crayfish burrows as retreat sites. She tracked 13 frogs to burrows, which ranged from 63–624 m from the center of their breeding wetlands, with an average of 315 m (Heemeyer's mean for our Northern Crawfish Frog populations was about 350 m).

Further, Chelsea Kross has been working on several aspects of Crawfish Frog population biology for her doctorate in the Department of Biological Sciences at the University of Arkansas Fayetteville. This information on Arkansas Crawfish Frog populations will be novel, and may help fill in gaps between our population, located at the extreme northeastern portion of the species' distribution, and the Attwater population, representing the extreme southwestern corner of the species' distribution.

Time It Was and What a Time It Was

20

The spring of 2009 was a heady time for the Lannoo Lab, which by then had been running steady for close to two decades. What began in 2009 was similar to the effort Priya Nanjappa, Laura Blackburn, and MJL had previously put into assembling *Amphibian Declines*, the massive book (225 contributors and "too big to fit in a backpack") created as a one-stop reference for the American amphibian conservation community. But unlike *Amphibian Declines*, which in essence was a scholarship/mapmaking exercise, the Crawfish Frog project was fully immersed in the long history of North American field biology. As MJL has noted (Lannoo 2018, p. xv):

> In its most general sense, field biology is simply studying nature in nature … because primary observations must come first, field biology will usually involve some sort of lifestyle disruption due to the need to be at that special place during that particular time when one can best, or perhaps only, attend to that specific subject. In short, field biology involves some level of self-sacrifice. There is a cost—travel, sleep, comfort, perhaps relationships. This tradition of self-sacrifice goes back a long way. In 1834, when Thomas Nuttall and John Kirk Townsend were collecting specimens of never-before-described plants, birds, and mammals during the Wyeth expedition to the Oregon country, they placed the safety of their collections over personal comfort and well-being. Shortly after beginning their expedition, Nuttall wrote, "Already we have cast away all our useless and superfluous clothing and have been content to mortify our natural pride, to make room for our specimens."

After consultations with John Crawford, Bill Peterman, and others outside our lab, and following an international search, MJL brought on Nate Engbrecht, Jen Heemeyer, and Vanessa Kinney to run aspects of the Crawfish Frog project, as detailed above, for their master's degrees. After their graduations in late 2010 and early 2011, Heemeyer left, Kinney stayed on for another six months, and Engbrecht stayed on for another two years. During the summer of 2011, MJL hired Mike Sisson, a seasoned Gopher Frog researcher, to bridge the gap. Jaimie Klemish, a Ph.D. student on another Lannoo Lab project, also helped to bridge this gap. In 2012, MJL brought on RMS to work on the Crawfish Frog project as a Ph.D. student. In addition to her project, RMS helped to supervise Jonathan Swan's master's project. Andrew Hoffman assisted with raising Crawfish Frog tadpoles.

We also enjoyed some remarkable technicians, including Jagger Foster, John Ryan, Austin May, Tenia Wheat, Alex Robinson, Kelly Robinson (no relation), Leisha Neumann, Lauren Sawyer, Helen Nesius, Alana Whitlock, Jessica Clevenger, Michael Goode, Suzie Ronk, Susie Lannoo, and Pete Lannoo (Appendix II). We collaborated with other research labs, including those of Drs. John Whitaker, the late Daryl Karns, Joe Robb, Allan Pessier, Stephen Richter, the late Marcy Sieggren, Irene Macallister, Alisa Gallant, Robert Klaver, Dan Saenz, and Toby Hibbitts.

Fieldwork takes some getting used to, especially fieldwork conducted at night in the rain. But we had a critical issue we needed to address separate from the discomforts of sleep deprivation and always being soaked to the skin. Because of our consistent nocturnal activity patterns, alarmed neighbors called the cops. After some initial discomfort at being approached and questioned by the local conservation officers (COs) and sheriff's deputies, we got to know and befriended them. The COs informed us that our field site was in a notorious methamphetamine area, and if we saw anything suspicious we should turn off our headlamps, drop into the prairie grass, and call them. MJL and one of our female technicians decided to not take any chances, and began carrying sidearms. At night, we always worked in groups, although on nights where we anticipated light frog activity (too cold or too dry) MJL usually went out alone. Our caution was well founded. On 6 June, 2013, two meth heads from the town just north of our study site killed 19-year-old Katelyn Wolfe at our study site; at the strip-pit lake next to New Pond. We were not out there at the time, but, like everyone who heard about this violent and senseless crime, we were deeply shaken. In planning our study, we tried to take into account every ecological variable we could imagine. Initially, however, we didn't take into account—because we couldn't imagine—the dark underbelly of the human culture where we worked.

On the other hand, once we introduced ourselves and explained what we were doing, the families living around our study site couldn't have been nicer. They stopped by to chat, loaned us raccoon traps, and offered us shelter during thunderstorms. We were surprised that their kindness toward us did not always extend to the DNR. On 19 September, 2009, not long after we'd started our project, Ron Ronk, then the DNR property manager, and his crew burned two sections of our study site (this was a normal management practice that we were pleased to discover did not directly impact Crawfish Frogs). As Vanessa and Jen were standing on the side of the road, watching the fire, an old rusty pickup truck pulled up. A scruffy guy put his elbow on the doorsill, stuck his head out, spit like Clint Eastwood, and grumbled, referring to the DNR, "Them bastards burn your frogs?" While we were not surprised at the anti-government senti-ment, we had never seen this person before and Vanessa and Jen thought it was creepy that he knew so much about what we were doing. In a similar vein, the next spring Jen found she had a suitor. One of the locals stopped by to chat, then began leaving morel mushrooms on the hood of her car. She discouraged him as best she could, and until he stopped, we made sure she was never out alone.

Every morning, after either being up all night or just out at daybreak, before heading back to the lab, we'd stop at the Casey's store in Jasonville for a coffee and a chocolate-iced donut. We were usually muddy and tired, and the caffeine/sugar pick-me-up was just enough to cover the 45-minute drive home. The mayor's wife, Nancy, worked the counter, and after a few weeks she began looking past our appearance and odor, and befriended us; she sort of adopted Vanessa, Jen, and Nate, and later Rochelle. Every morning she would ask them about their frogs, and they would ask her about her grandkids. It was a cheery connection with this little town, and after Nancy retired we discovered that the donuts didn't taste quite as good.

The most underappreciated piece of equipment we had was our cell phones. We would, of course, use them to text or call while we were in the field, or when some of us were in the field and others were at home or in the lab with access to a computer with our pit tag records. And we would use our cell phones to take photos of animals or questionable toe clips. But the feature we came to depend on the most was the weather app. During the breeding sea-son, Crawfish Frog activity was often triggered by nighttime thunderstorms. But because of the danger posed by lightning on an open prairie (where we were the tallest objects around), we could not work during the storms. So, at home, we would set our alarms for when storms were predicted to roll though, get up, and drive down to HFWA-W through the lightning, thunder, rain, and sometimes hail (Jen reminded us that we often sought shelter from hail under the canopy of the funeral home in Jasonville). We would park and wait for the storms to clear, so we could start catching and processing frogs. One night in 2015, after a thunderhead rumbled off to the east, still flashing lightning, a full moon emerged from behind the tall cloud bank. It was so bright we didn't need headlamps to follow the path to the wetlands, so we turned them off. Leisha Neumann was walking with MJL, and then she wasn't. When he turned around to ask her if she was all right, she was standing still, looking up, and said, "This is the most beautiful sky I've ever seen."

Without a doubt one of the most difficult aspects of this project for the graduate students and technicians was learning to get comfortable working at night in the rain. But they all did. Another thing they had to learn, and this was admittedly unfortunate, was to get used to seeing animals as data. During the field seasons of 2009 and 2010, the only two years when we kept track, we processed just shy of 19,000 amphibians and reptiles, mostly at our drift-fences. That's a lot of processing. And because we always seemed to be time limited, we were necessarily all business when handling animals; and being all business, we lost our sense of wonder about them. But every so often in the middle of the night, soaked and muddy, tired and achy, we would hear the distant call of a Crawfish Frog; out there, lonely and plaintive, and we would stop what we were doing, look at each other, and smile.

Appendix I: Listing of Amphibian and Reptile Species Detected at Our Study Site, Hillenbrand Fish and Wildlife Area-West (Terrell et al. 2014b)

Salamanders (5)

Ambystoma opacum	Marbled Salamander
Ambystoma texanum	Small-Mouthed Salamander
Ambystoma tigrinum	Eastern Tiger Salamander
Notophthalmus viridescens	Eastern Newt
Plethodon cinereus	Eastern Red-Backed Salamander

Frogs (9)

Acris blanchardi	Blanchard's Cricket Frog
Bufo fowleri	Fowler's Toad
Hyla chrysoscelis	Cope's Gray Treefrog
Pseudacris crucifer	Spring Peeper
Pseudacris triseriata	Western Chorus Frog
Rana areolata	Crawfish Frog
Rana catesbeiana	American Bullfrog
Rana clamitans	Green Frog
Rana sphenocephala	Southern Leopard Frog

Snakes (14)

Carphophis amoenus	Common Wormsnake
Clonophis kirtlandii	Kirtland's Snake
Coluber constrictor	North American Racer
Diadophis punctatus	Ring-Necked Snake
Heterodon platirhinos	Eastern Hog-Nosed Snake
Lampropeltis calligaster	Prairie Kingsnake
Lampropeltis getula	Eastern Kingsnake
Lampropeltis triangulum	Eastern Milksnake
Nerodia sipedon	Common Watersnake
Opheodrys aestivus	Rough Greensnake
Pantheropis obsoletus	Western Ratsnake
Storeria dekayi	Dekay's Brownsnake
Thamnophis saurita	Eastern Ribbonsnake
Thamnophis sirtalis	Common Gartersnake

Turtles (5)

Chelydra serpentina	Snapping Turtle
Chrysemys picta	Painted Turtle
Sternotherus odoratus	Eastern Musk Turtle
Terrapene carolina	Eastern Box Turtle
Trachemys scripta	Pond Slider

Lizards (2)

Plestiodon fasciatus	Common Five-Lined Skink
Scincella lateralis	Little Brown Skink

Appendix II: Personnel

David Bakken, Technician
Dr. John Crawford, Postdoctoral Fellow
Nate Engbrecht, Indiana State University, M.Sc.
B. Jagger Foster, Technician
Dr. Alisa Gallant, U.S. Geological Survey
Todd Gerardot, U.S. Fish & Wildlife Service
Michael Goode, Technician
Emily Gustin, Eastern Kentucky University
Dr. Tim Halliday, The Open University (Deceased)
Jennifer Heemeyer, Indiana State University, M.Sc.
Dr. Toby Hibbitts, Texas A&M University
Andrew Hoffman, Indiana State University, Graduate Student
Dr. Daryl Karns, Hanover College (Deceased)
Vanessa Kinney, Indiana State University, M.Sc.
Dr. Robert Klaver, U.S. Geological Survey
Dr. Michael Lannoo, Indiana University School of Medicine
Peter Lannoo, Technician
Susan Lannoo, Consultant
Dr. Irene Macallister, U.S. Army Corps of Engineers
Austin May, Technician
Austin McClain, Technician
Helen Nesius, Technician
Leisha Neumann, Technician
Harrison Ndife, Technician
Dr. Allan Pessier, San Diego Zoo/Washington State University
Dr. William Peterman, Graduate Student
Wyatt Pommier, Technician
Dr. Stephen Richter, Eastern Kentucky University
Dr. Joe Robb, U.S. Fish & Wildlife Service
Alex Robinson, Technician
Kelly Robinson, Technician, M.Sc.
Suzie Ronk, Technician
John Ryan, Technician
Dr. Dan Saenz, U.S. Forest Service
Dr. Lauren Sawyer, Technician

Dr. Danny Schaefer, Technician
Marcy Sieggreen, Ph.D. candidate, Detroit Zoological Society (Deceased)
Shane Stephens, Technician
Dr. Rochelle Stiles, Indiana State University, Ph.D.
Jonathan Swan, Indiana State University, M.Sc.
Ben Walker, U.S. Fish & Wildlife Service
Tenia Wheat, Technician
Dr. John Whitaker, Indiana State University
Alana Whitlock, Technician
Dr. Perry Williams, U.S. Fish & Wildlife Service

Appendix III: Lannoo Lab Crawfish Frog Products

2009

Lannoo, M. J., V. C. Kinney, J. L. Heemeyer, N. J. Engbrecht, A. L. Gallant, and R. W. Klaver. 2009. Mine spoil prairies expand critical habitat for endangered and threatened amphibian and reptile species. *Diversity* 1:118–132.

2010

Engbrecht, N. J. 2010. The status of Crawfish Frogs (*Lithobates areolatus*) in Indiana, and a tool to assess populations. MS thesis, Indiana State Univ.

Engbrecht, N. J., and J. L. Heemeyer. 2010. *Lithobates areolatus circulosus*: predation. *Herpetological Review* 41:197.

Engbrecht, N. J., and M. J. Lannoo. 2010. A review of the status and distribution of Crawfish Frogs (*Lithobates areolatus*) in Indiana. *Proceedings of the Indiana Academy of Science* 119:64–73.

Heemeyer, J. L., V. C. Kinney, N. J. Engbrecht, and M. J. Lannoo. 2010. The biology of Crawfish Frogs (*Lithobates areolatus*) prevents the full use of telemetry and drift fence techniques. *Herpetological Review* 41:42–45.

Heemeyer, J. L., and M. J. Lannoo. 2010. A new technique for capturing burrow-dwelling anurans. *Herpetological Review* 41:168–170.

Heemeyer, J. L., J. G. Palis, and M. J. Lannoo. 2010. *Lithobates areolatus circulosus* (Northern Crawfish Frog): predation. *Herpetological Review* 41:475.

Hoffman, A. S., J. L. Heemeyer, P. J. Williams, J. R. Robb, D. R. Karns, V. C. Kinney, N. J. Engbrecht, and M. J. Lannoo. 2010. Strong site fidelity and a variety of imaging techniques reveal around-the-clock and extended activity patterns in Crawfish Frogs (*Lithobates areolatus*). *BioScience* 60:829–834.

Kinney, V. C., N. J. Engbrecht, J. L. Heemeyer, and M. J. Lannoo. 2010. New county records for amphibians and reptiles in southwest Indiana. *Herpetological Review* 41:387.

Kinney, V. C., and M. J. Lannoo. 2010. *Lithobates areolatus circulosus* (Northern Crawfish Frog): breeding. *Herpetological Review* 41:197–198.

2011

Engbrecht, N. J., S. J. Lannoo, J. O. Whitaker, and M. J. Lannoo. 2011. Comparative morphometrics in ranid frogs (subgenus *Nenirana*): are apomorphic elongation and a blunt snout responses to deep, small-bore burrow dwelling in Crawfish Frogs (*Lithobates areolatus*)? *Copeia* 2011:285–295.

Heemeyer, J. L. 2011. Breeding migrations, survivorship, and obligate crayfish burrow use by adult Crawfish Frogs (*Lithobates areolatus*). MS thesis, Indiana State Univ.

Heemeyer, J. L., and M. J. Lannoo. 2011. *Lithobates areolatus circulosus* (Northern Crawfish Frog): winterkill. *Herpetological Review* 42:261–262.

Kinney, V. C. 2011. Adult survivorship and juvenile recruitment in populations of Crawfish Frogs (*Lithobates areolatus*), with additional consideration of the population sizes of associated pond breeding species. MS thesis, Indiana State Univ.

Kinney, V. C., J. L. Heemeyer, A. P. Pessier, and M. J. Lannoo. 2011. Seasonal pattern of *Batrachochytrium dendrobatidis* infection and mortality in *Lithobates areolatus*: affirmation of Vredenburg's "10,000 zoospore rule." *PLoS One* 6(3). doi:10.1371/journal.pone.0016708.

2012

Engbrecht, N. J, J. L. Heemeyer, V. C. Kinney, and M. J. Lannoo. 2012. *Lithobates areolatus* (Crawfish Frogs): thwarted predation. *Herpetological Review* 43:323–324.

Engbrecht, N. J., and M. J. Lannoo. 2012. Crawfish Frog behavioral differences in postburned and vegetated grasslands. *Fire Ecology* 8:63–76.

Heemeyer, J. L., and M. J. Lannoo. 2012. Breeding migrations in Crawfish Frogs (*Lithobates areolatus*): long-distance movements, burrow philopatry, and mortality in a near-threatened species. *Copeia* 2012:440–450.

Heemeyer, J. L., P. J. Williams, and M. J. Lannoo. 2012. Obligate crayfish burrow use and core habitat requirements of Crawfish Frogs. *Journal of Wildlife Management* 76:1081–1091.

2013

Engbrecht, N. J., P. J. Williams, J. R. Robb, D. R. Karns, M. J. Lodato, T. A. Gerardot, and M. J. Lannoo. 2013. Is there hope for the Hoosier Frog? An update on the status of Crawfish Frogs (*Lithobates areolatus*) in Indiana, with recommendations for their conservation. *Proceedings of the Indiana Academy of Science* 121:147–157.

Klemish, J. L., N. J. Engbrecht, and M. J. Lannoo. 2013. Positioning minnow traps in wetlands to avoid accidental deaths of frogs. *Herpetological Review* 44:241–242.

Nunziata, S. O., M. J. Lannoo, J. R. Robb, D. R. Karns, S. L. Lance, and S. C. Richter. 2013. Population and conservation genetics of Crawfish Frogs, *Lithobates areolatus*, at the northeastern range limit. *Journal of Herpetology* 47:361–368.

Williams, P. J., N. J. Engbrecht, J. R. Robb, V. C. K. Terrell, A. P. Pessier, and M. J. Lannoo. 2013. Surveying a threatened species through a narrow detection window. *Copeia* 2013:553–562.

2014

Terrell, V. C. K., N. J. Engbrecht, A. P. Pessier, and M. J. Lannoo. 2014. Drought reduces chytrid fungus (*Batrachochytrium dendrobatidis*) infection intensity and mortality but not prevalence in adult Crawfish Frogs (*Lithobates areolatus*). *Journal of Wildlife Diseases* 50:56–62.

Terrell, V. C. K., J. L. Klemish, N. J. Engbrecht, J. A. May, P. J. Lannoo, R. M. Stiles, and M. J. Lannoo. 2014. Amphibian and reptile colonization of reclaimed coal spoil grasslands. *Journal of North American Herpetology* 2014:59–68.

2015

Engbrecht, N. J., J. L. Heemeyer, C. G. Murphy, R. M. Stiles, J. W. Swan, and M. J. Lannoo. 2015. Upland calling behavior in Crawfish Frogs (*Lithobates areolatus*) and calling triggers caused by noise pollution. *Copeia* 103:1048–1057.

2016

Stiles, R. M. 2016. Amphibian response to wetlands in restored and reclaimed grass-lands: passive recolonization and active reintroduction techniques. Ph.D. diss., Indiana State Univ.

Stiles, R. M., M. Sieggreen, A. Preston, A. P. Pessier, S. J. Lannoo, and M. J. Lannoo. 2016b. First report of ranavirus-associated mortality in Crawfish Frogs (*Lithobates areolatus*), a species of conservation concern, in Indiana, USA. *Herpetological Review* 47:389–391.

Stiles, R. M., M. J. Sieggreen, R. A. Johnson, K. Pratt, M. Vassallo, M. Andrus, M. Perry, J. W. Swan, and M. J. Lannoo. 2016c. Captive-rearing state endangered Crawfish Frogs *Lithobates areolatus* from Indiana, USA. *Conservation Evidence* 13:7–11.

Stiles, R. M, J. W. Swan, J. L. Klemish, and M. J. Lannoo. 2016a. Amphibian habitat creation on postindustrial landscapes: a case study in a reclaimed coal strip-mine area. *Canadian Journal of Zoology*. doi:10.1139/cjz-2015-0163.

Swan, J. W. 2016. Artificial burrow use by juvenile Crawfish Frogs (*Lithobates areolatus*). MS thesis, Indiana State Univ.

2017

Lannoo, M. J., and R. M. Stiles. 2017. Effects of short-term climate variation on a long-lived frog. *Copeia* 105:726–733.

Lannoo, M. J., R. M. Stiles, M. A. Sisson, J. W. Swan, V. C. K. Terrell, and K. E. Robinson. 2017. Patch dynamics inform management decisions in a threatened frog species. *Copeia* 105:53–63.

Stiles, R. M., T. R. Halliday, N. J. Engbrecht, J. W. Swan, and M. J. Lannoo. 2017. Wildlife cameras reveal high resolution activity patterns in threatened Crawfish Frogs (*Lithobates areolatus*). *Herpetological Conservation Biology* 12:160–170.

2018

Lannoo, M. J., R. M. Stiles, D. Saenz, and T. J. Hibbitts. 2018. Comparative call characteristics in the anuran subgenus *Nenirana*. *Copeia* 106:575–579.

2020

Stiles, R. M., V. C. K. Terrell, J. C. Maerz, and M. J. Lannoo. Survivorship, fitness, and carry-over effects in Crawfish Frogs (*Rana areolata*), a species of conservation concern. *Copeia* (in press).

Terrell, V. C. K., J. C. Maerz, N. J. Engbrecht, R. M. Stiles, B. A. Crawford, and M. J. Lannoo. Population dynamics of threatened Crawfish Frogs informs targets for management. *Copeia* (in press).

We have deposited our Crawfish Frog specimens at the Field Museum of Natural History (FMNH), catalog numbers 288547–288590. We have deposited our recordings of Northern Crawfish Frog breeding calls at the Cornell Laboratory of Ornithology's Macaulay Library, catalog numbers 230118–230122. We have uploaded a video of Romeo upland calling: https://www.youtube.com/watch?time_continue=3&v=ojO5BrCEfsU

Acknowledgments

We thank K. Gremillion-Smith, S. Klueh-Mundy, S. Johnson, R. Ronk, B. Feaster, and T. Stoelting of the Indiana Department of Natural Resources for supporting this research and for allowing us to work at HFWA-W. We especially thank R. Ronk for allowing us to work at HFWA-W, his patience in dealing with our concerns, and his friendship. This work was funded by State Wildlife Grant E2-08-WDS13 through the Indiana Department of Natural Resources, Division of Fish and Wildlife; the National Amphibian Conservation Center at the Detroit Zoological Society; an Amphibian Specialists Group and Amphibian Research and Monitoring Initiative Seed Grant; an Amphibian Taxon Advisory Group Small Grant; Eastern Kentucky University's Department of Biological Sciences; Department of Energy award DE-FC09-07SR22506 to the University of Georgia Research Foundation; Texas Parks and Wildlife Department Wildlife Conservation Grant CFDA #15.634; and a research grant awarded by Prairie Biotic Research, Inc. Dr. Peter Duong at Indiana University School of Medicine–Terre Haute and Peabody Energy provided critical student fellowships.

This work was conducted under Indiana State University's Animal Care and Use (IACUC) numbers 245168-1: ML; 445698-1:ML, 3-24-2008, 628051-1:ML, and 705946-1:ML; Indiana Scientific Collection (DNR) permits 09-0112, 10-0027, 11-0017, 12-0015, 13-0072, 14-063, and 15-013, and 16-058; and Indiana Department of Natural Resources Importation Permit numbers 0162, 0618, and 0171.

A special thanks to the following biologists: P. Moler and D. Scott for collecting belostomatids; C. K. Adams, A. Bryant, J. Childress, and E. Childress of the U.S. Forest Service, and L. Smith of the Joseph W. Jones Ecological Research Center at Ichaway in Newton, Georgia, for sending us recordings of advertisement calls; T. J. LaDuc at the Texas Natural History Collection at the University of Texas, Austin, and M. A. Kwiatkowski of the Department of Biology Vertebrate Museum at Stephen F. Austin State University for access to their specimens; the National Amphibian Conservation Center (Detroit Zoological Society) staff, B. Johnson, K. Pratt, M. Vassallo, M. Andrus, and M. Perry for raising and transporting Crawfish Frog tadpoles; J. Eastman for a consultation on radiographic distortion; J. Humphries for sharing his observations on Gopher Frog burrowing, and all the participants in the 2010 SEPARC Workshop for sharing their insights and commenting on our distribution map;

L. Weir for a discussion on amphibian distributions; X. Bernal for discussions on anuran calling; D. McGowan at Ravenswood Media for filming and publicizing aspects of our project; and J. McGowan, M. Medler, M. Webster, and M. A. Young at Cornell University's Laboratory of Ornithology for allowing us to deposit our recordings of Crawfish Frogs. The following people read and commented on earlier drafts of this book: N. Engbrecht, P. Eyheralde, J. Heemeyer, V. Kinney Terrell, S. Lannoo, R. Ronk, L. Smith, and S. Walls.

In addition to the biologists mentioned above, and the personnel mentioned in Appendix II, we thank the following individuals for their assistance on or during various aspects of the projects summarized here: M. Angilletta, T. Anton, R. Arndt, B. Becker, E. Brinson, J. Brinson, P. Cain, G. Casper, B. Donahue, P. Doung, R. Drewes, D. Dugas, T. Feaster, S. Galatowitsch, M. Greenan, J. Hanken, R. Hedge, R. Hellmich, D. Hews, T. Hibbitts, K. Krysko, M. Lang, J. Latimer, A. Leffel, S. Lima, N. Lindel, J. Losos, D. Lowe, J. MacGregor, J. Maerz, T. Mann, D. McGowan, V. Meretsky, R. Millar, W. Mitchell, B. Nesius, M. Nickerson, R. Nussbaum, C. Phillips, J. Pruett, D. Ramesh, A. Resetar, S. Richter, S. Rogers, D. Ronk, J. Rosado, G. Schneider, F. Scott, P. Scott, A. Sherman, L. Sterrenburg, G. Stillings, R. Stoelting, R. Stonebraker, Z. Terrell, A. Timm, Z. Walker, J. Weber, J. Whitaker, and R. Williams.

References

Aihara, I., T. Mizumoto, T. Otsuka, H. Awano, K. Nagira, H. G. Okuno, and K. Aihara. 2014. Spatio-temporal dynamics in collective frog choruses examined by mathematical modeling and field observations. *Scientific Reports* 4. doi:10.1038/srep03891.

Alberch, P. 1986. Possible dogs. *Natural History* 95(12):4–8.

Alford, R. A., and H. M. Wilbur. 1985. Priority effects in experimental pond communities: competition between *Bufo* and *Rana*. *Ecology* 66:1097–1105.

Ali, M. F., K. R. Lips, F. C. Knoop, B. Fritzsch, C. Miller, and J. M. Conlon. 2002. Antimicrobial peptides and protease inhibitors in the skin secretions of the Crawfish Frog, *Rana areolata*. *Biochim Biophys Acta* 1601:55–63.

Altig, R. 1972. Defensive behavior in *Rana areolata* and *Hyla avivoca*. *Quarterly Journal of the Florida Academy of Sciences* 35:212–216.

Altig, R., and R. Lohoefener. 1983. *Rana areolata* Baird and Girard Crawfish Frog. *Catalogue of American Amphibians and Reptiles* 324.1–324.4.

Altwegg, R. 2003. Multistage density dependence in an amphibian. *Oecologia* 136:46–50.

Alverez, D., and A. G. Nicieza. 2002. Effects of induced variation in anuran larval development on postmetamorphic energy reserves and locomotion. *Oecologia* 131:186–195.

Arnold, S. J., and R. J. Wassersug. 1978. Differential predation on metamorphic anurans by Garter Snakes (*Thamnophis*): social behavior as a possible defense. *Ecology* 59:1014–1022.

Bailey, R. M. 1943. Four species new to the Iowa herpetofauna, with notes on their natural histories. *Proceedings of the Iowa Academy of Science* 50:347–352.

Baird, S. F., and C. F. Girard. 1852. Characteristics of some new reptiles in the Museum of the Smithsonian Institution, Washington, D.C. *Proceedings of the Academy of Natural Sciences of Philadelphia* 6:173.

Barbour, R. B. 1971. *Amphibians & reptiles of Kentucky*. Lexington, KY: University Press of Kentucky.

Beck, C. W., and J. D. Congdon. 2000. Effects of age and size at metamorphosis on growth and survivorship of Southern Toad (*Bufo terrestris*) metamorphs. *Canadian Journal of Zoology* 77:944–951.

Berger, W. H. 2010. A. E. Douglas (1867–1962) and solar cycles in tree rings. Scripps Institute of Technology Technical Report, La Jolla, California. http://escholarship. org/uc/item/7gw6m710 (accessed 13 November 2017).

Berven, K. A. 1990. Factors affecting population fluctuations in larval and adult stages of the Wood Frog (*Rana sylvatica*). *Ecology* 71:1599–1608.

Bissell, T. 2001. The last lion. *Outside Magazine*. https://www.outsideonline. com/1893296/last-lion.

Blatchley, W. S. 1900. Notes on the batrachians and reptiles of Vigo County, Indiana (II). *Annual Report to the Indiana Department of Geology and Natural Resources* 24:537–552.

Bossuyt, F., R. M. Brown, D. M. Hillis, D. C. Canatella, and M. C. Milinkovitch. 2006. Phylogeny and biogeography of a cosmopolitan frog radiation: late Cretaceous diversification resulted in continental-scale endemism in family Ranidae. *Systematic Biology* 55:579–594.

Boulenger, G. A. 1920. A monograph of the American frogs in the genus *Rana*. *Proceedings of the American Academy of Arts and Sciences* 55:413–480.

Box, G. E. P., and R. Draper. 1987. *Empirical model-building and response surfaces*. New York, NY: John Wiley and Sons.

Bradford, D. F. 2005. Factors implicated in amphibian population declines in the United States. In *Amphibian declines: the conservation status of United States species*, ed. M. J. Lannoo, 915–925. Berkeley, CA: University of California Press.

Bragg, A. N. 1953. A study of *Rana areolata* in Oklahoma. *The Wasmann Journal of Biology* 11:273–318.

Briggs, C. J., R. A. Knapp, and V. T. Vredenburg. 2010. Enzootic and epizootic dynamics of the chytrid fungal pathogen of amphibians. *Proceedings of the National Academy of Sciences of the United States of America* 107:9695–9700.

Brown, L. E., H. O. Jackson, and J. R. Brown. 1972. Burrowing behavior of the Chorus Frog, *Pseudacris streckeri*. *Herpetologica* 28:325–328.

Busby, W. H., and W. R. Brecheisen. 1997. Chorusing phenology and habitat associations of the Crawfish Frog, *Rana areolata* (Anura: Ranidae), in Kansas. *Southwestern Naturalist* 42:210–217.

Butler, A. B., and W. Hodos. 1996. *Comparative vertebrate neuroanatomy: evolution and adaptation*. New York, NY: Wiley-Liss, Inc.

Cabrera-Guzmán, E., M. R. Crossland, G. P. Brown, and R. Shine. 2013. Larger body size at metamorphosis enhances survival, growth and performance of young Cane Toads (*Rhinella marina*). *PLoS One* 8:e70121.

Cagle, F. R. 1942. Herpetological fauna of Jackson and Union Counties, Illinois. *American Midland Naturalist* 28:164–200.

Carr, A. F., Jr. 1940. *Herpetology of Florida*. University of Florida Publication, Biological Science Series. Volume III, Number 1. Gainesville, FL: University of Florida.

Chadwick, D. H. 2010. *The Wolverine way*. Ventura, CA: Patagonia Books.

Chan, W.-P., I.-C. Chen, R. K. Colwell, W.-C. Liu, C.-Y. Huang, and S.-F. Shen. 2016. Seasonal and daily climate variation have opposite effects on species elevational range size. *Science* 351:1437–1439.

Chelgren, N. D., D. K. Rosenberg, S. S. Heppell, and A. I. Gitelman. 2006. Carryover aquatic effects on survival of metamorphic frogs during pond emigration. *Ecological Applications* 16:627–636.

Cochran, D. M. 1961. *Type specimens of reptiles and amphibians in the U.S. National Museum*. Washington, DC: Smithsonian Institution Press.

Collins, J. P., and A. Storfer. 2003. Global amphibian declines: sorting the hypotheses. *Diversity and Distributions* 9:89–98.

Conant, R. 1958. *A field guide to reptiles and amphibians of the United States and Canada east of the 100th meridian*. Boston, MA: Houghton Mifflin Company.

Conant, R., and J. T. Collins. 1998. *A field guide to reptiles and amphibians of eastern and central North America*, 3rd ed. Boston, MA: Houghton Mifflin Company.

Cope, E. D. 1875. Check-list of North American Batrachia and Reptilia; with a systematic list of the higher groups, and an essay on geographical distribution. Based on specimens contained in the United States National Museum. *United States National Museum Bulletin* 1:1–104.

Corn, P. S. 2003. Amphibian breeding and climate change: importance of snow in the mountains. *Conservation Biology* 17:622–625.

Daszak, P., D. E. Scott, A. M. Kilpatrick, C. Faggioni, J. W. Gibbons, and D. Porter. 2005. Amphibian population declines at Savannah River site are linked to climate, not chytridiomycosis. *Ecology* 86:3232–3237.

Davies, M. 1984. Osteology of the myobatrachine frog *Arenophryne rotunda* Tyler (Anura: Leptodactylidae) and comparisons with other myobatrachine genera. *Australian Journal of Zoology* 32:789–802.

Dawkins, R., and J. Krebs. 1978. Chapter 10: animal signals: information or manipulation? In *Behavioural ecology: an evolutionary approach*, ed. J. R. Krebs and N. B. Davies, 282–309. Sunderland, MA: Sinauer Associates Inc.

Dickerson, M. C. 1906. *The frog book*. Ithaca, NY: Comstock Publ. Co.

Diego-Rasilla, F. J., R. M. Luengo, and J. B. Phillips. 2008. Use of a magnetic compass for nocturnal orientation in the Palmate Newt, *Lissotriton helveticus*. *Ethology* 114:808–815.

Diffenbaugh, N. S., J. S. Pal, R. J. Trapp, and F. Giorgi. 2005. Fine-scale processes regulate the response of extreme events to global climate change. *Proceedings of the National Academy of Sciences of the United States of America* 102:15774–15778.

Dowling, H. G. 1957. A review of the amphibians and reptiles of Arkansas. Occasional Papers of the University of Arkansas Museum, Fayetteville, AR.

Dundee, H. A., and D. A. Rossman. 1989. *The amphibians and reptiles of Louisiana*. Baton Rouge, LA: Louisiana State University Press.

Earl, J. E., and R. D. Semlitsch. 2013. Carryover effects in amphibians: are characteristics of the larval habitat needed to predict juvenile survival? *Ecological Applications* 23:1429–1442.

Earl, J. E., and H. H. Whiteman. 2015. Are commonly used fitness predictors accurate? A meta-analysis of amphibian size and age at metamorphosis. *Copeia* 103:297–309.

Edwards, B. A., D. A. Jackson, and K. M. Somers. 2009. Multispecies crayfish declines in lakes: implications for species distributions and richness. *Journal of the North American Benthological Society* 28:719–732.

Emerson, S. 1976. Burrowing in frogs. *Journal of Morphology* 149:437–458.

Engbrecht, N. J., and J. L. Heemeyer. 2010. *Lithobates areolatus circulosus*. Predation. *Herpetological Review* 41:197.

Engbrecht, N. J., J. L. Heemeyer, V. C. Kinney, and M. J. Lannoo. 2012. *Lithobates areolatus* (Crawfish Frogs): thwarted predation. *Herpetological Review* 43:323–324.

Engbrecht, N. J., J. L. Heemeyer, C. G. Murphy, R. M. Stiles, J. W. Swan, and M. J. Lannoo. 2015. Upland calling behavior in Crawfish Frogs (*Lithobates areolatus*) and calling triggers caused by noise pollution. *Copeia* 103:1048–1057.

Engbrecht, N. J., and M. J. Lannoo. 2010. A review of the status and distribution of Crawfish Frogs (*Lithobates areolatus*) in Indiana. *Proceedings of the Indiana Academy of Science* 119:64–73.

Engbrecht, N. J., and M. J. Lannoo. 2012. Crawfish Frog behavioral differences in postburned and vegetated grasslands. *Fire Ecology* 8:63–76.

Engbrecht, N. J., S. J. Lannoo, J. O. Whitaker, and M. J. Lannoo. 2011. Comparative morphometrics in ranid frogs (subgenus *Nenirana*): are apomorphic elongation and a blunt snout responses to small-bore burrow dwelling in Crawfish Frogs (*Lithobates areolatus*)? *Copeia* 2011:285–295.

Engbrecht, N. J., P. J. Williams, J. R. Robb, D. R. Karns, M. J. Lodato, T. A. Gerardot, and M. J. Lannoo. 2013. Is there hope for the Hoosier Frog? An update on the status of Crawfish Frogs (*Lithobates areolatus*) in Indiana, with recommendations for their conservation. *Proceedings of the Indiana Academy of Science* 121:147–157.

Environmental Protection Agency (EPA). 2016. Climate change: midwest. https://www3.epa.gov/climtechange/impacts/midwest.html (accessed 14 February 2017).

Fauth, J. E. 1990. Interactive effects of predators and early larval dynamics of the treefrog *Hyla chrysoscelis*. *Ecology* 71:1609–1616.

Ferreira, C., and C. A. Ríos-Saldaña. 2016. Hail local fieldwork, not just global models. *Nature* 534:326.

Fortey, R. 2008. *Dry storeroom no. 1: the secret life of the natural history museum*. New York, N.Y.: Harper Press.

Gallant, A. L., R. W. Klaver, G. S. Casper, and M. J. Lannoo. 2007. Global rates of habitat loss and implications for amphibian conservation. *Copeia* 2007:965–977.

Gerhardt, H. C. 1975. Sound pressure levels and radiation patterns of the vocalizations of some North American frogs and toads. *Journal of Comparative Physiology* 102:1–12.

Gerhardt, H. C. 1994. The evolution of vocalization in frogs and toads. *Annual Review of Ecology and Systematics* 25:293–324.

Gibbons, J. W., and D. H. Bennett. 1974. Determination of anuran terrestrial activity patterns by drift fence method. *Copeia* 1974:236–242.

Gloyd, H. K. 1928. Amphibians and reptiles of Franklin County. *Transactions of the Kansas Academy of Science* 31:116–119.

Goater, C. P. 1994. Growth and survival of postmetamorphic toads: interactions among larval history, density, and parasitism. *Ecology* 75:2264–2274.

Goin, C. J., and M. G. Netting. 1940. Art. VIII. A new Gopher Frog from the Gulf Coast, with comments upon the *Rana areolata* group. *Annals of the Carnegie Museum* 28:137–167, with two figures. Pittsburgh.

Gomez-Mestre, I., S. Kulkarni, and D. R. Buchholz. 2013. Mechanisms and consequences of developmental acceleration in tadpoles responding to pond drying. *PLoS One* 8(12). doi:10.1371/joiurnal.pone.0084266.

Green, D. M. 2003. The ecology of extinction: population fluctuation and decline in amphibians. *Biological Conservation* 111:331–343.

Greene, H. W. 2013. *Tracks and shadows: field biology as art*. Berkeley, CA: University of California Press.

Gunderson, A. R., and J. H. Stillman. 2015. Plasticity in thermal tolerance has limited potential to buffer ectotherms from global warming. *Proceedings of the Royal Society B* 282. doi:10.1098/rspb.2015.0401.

Gunderson, A. R., B. Tsukimura, and J. H. Stilman. 2017. Indirect effects of global change: from physiological and behavioral mechanisms to ecological consequences. *Integrative and Comparative Biology* 57:48–54.

Hanski, I. 1998. Metapopulation dynamics. *Nature* 396:41–49.

Harper, F. 1935. The name of the Gopher Frog. *Proceedings of the Biological Society of Washington* 48:79–82.

Harrison, X. A., J. D. Blount, R. Inger, D. R. Norris, and S. Bearhop. 2011. Carry-over effects as drivers of fitness differences in animals. *Journal of Animal Ecology* 80:4–18.

Haswell, E. 2004. Northern Crawfish Frogs (*Rana areolata circulosa*) at Big Oaks National Wildlife Refuge in southeastern Indiana. BS thesis, Hanover College.

Hauersperger, B. A. 2005. Breeding ecology of the Northern Crawfish Frog, *Rana areolata circulosa*. BS thesis, Hanover College.

Heemeyer, J. L., V. C. Kinney, N. J. Engbrecht, and M. J. Lannoo. 2010a. The biology of Crawfish Frogs (*Lithobates areolatus*) prevents the full use of telemetry and drift fence techniques. *Herpetological Review* 41:42–45.

Heemeyer, J. L., and M. J. Lannoo. 2010. A new technique for capturing burrow-dwelling anurans. *Herpetological Review* 41:168–170.

Heemeyer, J. L., and M. J. Lannoo. 2011. *Lithobates areolatus circulosus* (Northern Crawfish Frog): winterkill. *Herpetological Review* 42:261–262.

Heemeyer, J. L., and M. J. Lannoo. 2012. Breeding migrations in Crawfish Frogs (*Lithobates areolatus*): long-distance movements, burrow philopatry, and mortality in a near-threatened species. *Copeia* 2012:440–450.

Heemeyer, J. L., J. G. Palis, and M. J. Lannoo. 2010b. *Lithobates areolatus circulosus* (Northern Crawfish Frog): predation. *Herpetological Review* 41:475.

Heemeyer, J. L., P. J. Williams, and M. J. Lannoo. 2012. Obligate crayfish burrow use and core habitat requirements of Crawfish Frogs. *Journal of Wildlife Management* 76:1081–1091.

Hildebrand, M. 1974. *Analysis of vertebrate structure*. New York, NY: John Wiley and Sons.

Hillis, D. M., and T. P. Wilcox. 2005. Phylogeny of the new world true frogs (*Rana*). *Molecular Phylogenetics and Evolution* 34:299–314.

Hobson, J. A. 1967. Electrographic correlates of behavior in the frog with special reference to sleep. *Electroencephalography and Clinical Neurophysiology* 11:113–121.

Hoffman, A. S., J. L. Heemeyer, P. J. Williams, J. R. Robb, D. R. Karns, V. C. Kinney, N. J. Engbrecht, and M. J. Lannoo. 2010. Strong site fidelity and a variety of imaging techniques reveal around-the-clock and extended activity patterns in Crawfish Frogs (*Lithobates areolatus*). *BioScience* 60:829–834.

Holbrook, C. T., and J. W. Petranka. 2004. Ecological interactions between *Rana sylvatica* and *Ambystoma tigrinum*: evidence of interspecific competition and facultative intraguild predation. *Copeia* 2004:932–939.

Huntley, A. C., M. Donnelly, and H. B. Cohen. 1978. Sleep in the Western Toad (*Bufo boreas*). *Sleep Research* 14:141.

Hurter, J. 1911. *Rana areolata* Baird and Girard: Gopher Frog. In *Herpetology of Missouri*, Vol. 20, No. 5115–5117. Transactions of the Academy of Science of St. Louis.

Huxley, J. 1972. *Problems of relative growth* (reprint of 1932 ed.). New York, NY: Dover.

Inger, R. F. 1958. Comments on the definition of genera. *Evolution* 12:370–384.

John-Alder, H. B., and P. J. Morin. 1990. Effects of larval density on jumping ability and stamina in newly metamorphosed *Bufo woodhousii fowleri*. *Copeia* 1990:856–860.

Johnson, T. R. 2000. *The amphibians and reptiles of Missouri*, 2nd ed. Jefferson City, MO: Missouri Department of Conservation.

Jordan, D. S. 1878. *Manual of the vertebrates of the northern United States including the district east of the Mississippi River, and north of North Carolina and Tennessee, exclusive of marine species*, 2nd ed. Chicago, IL: Jansen, McClurg & Company.

Kinney, V. C. 2011. Adult survivorship and juvenile recruitment in populations of Crawfish Frogs (*Lithobates areolatus*), with additional consideration of the population sizes of associated pond breeding species. MS thesis, Indiana State University.

Kinney, V. C., N. J. Engbrecht, J. L. Heemeyer, and M. J. Lannoo. 2010. New county records for amphibians and reptiles in southwest Indiana. *Herpetological Review* 41:387.

Kinney, V. C., J. L. Heemeyer, A. P. Pessier, and M. J. Lannoo. 2011. Seasonal pattern of *Batrachochytrium dendrobatidis* infection and mortality in *Lithobates areolatus*: affirmation of Vredenburg's "10,000 zoospore rule." *PLoS ONE* 6(3):e16708. doi:10.1371/journal.pone.0016708.

Kinney, V. C., and M. J. Lannoo. 2010. *Lithobates areolatus circulosus* (Northern Crawfish Frog): breeding. *Herpetological Review* 41:197–198.

Klemish, J. L., N. J. Engbrecht, and M. J. Lannoo. 2013. Positioning minnow traps in wetlands to avoid accidental deaths of frogs. *Herpetological Review* 44:241–242.

Krebs, J. R., and N. B. Davies, eds. 1978. *Behavioural ecology: an evolutionary approach*. Sunderland, MA: Sinauer Associates Inc.

Kwiatkowski, M. A., D. Saenz, and T. Hibbitts. 2016. Habitat use and movement patterns of the Southern Crawfish Frog (*Rana areolata*). Unpublished report. Texas Parks and Wildlife Department, Austin, TX.

Lannoo, M. J., ed. 2005. *Amphibian declines: the conservation status of United States species*. Berkeley, CA: University of California Press.

Lannoo, M. J. 2018. *This land is your land: the story of field biology in America*. Chicago, IL: University of Chicago Press.

Lannoo, M. J., V. C. Kinney, J. L. Heemeyer, N. J. Engbrecht, A. L. Gallant, and R. W. Klaver. 2009. Mine spoil prairies expand critical habitat for endangered and threatened amphibian and reptile species. *Diversity* 1:118–132.

Lannoo, M. J., and R. M. Stiles. 2017. Effects of short-term climate variation on a long-lived frog. *Copeia* 105:726–733.

Lannoo, M. J., and R. M. Stiles. "Somebody's got to stay outside:" understanding shifting amphibian ecological relationships in a world of environmental change. *Herpetologica* (in press).

Lannoo, M. J., R. M. Stiles, D. Saenz, and T. J. Hibbitts. 2018. Comparative call characteristics in the anuran subgenus *Nenirana*. *Copeia* 106:575–579.

Lannoo, M. J., R. M. Stiles, M. A. Sisson, J. W. Swan, V. C. K. Terrell, and K. E. Robinson. 2017. Patch dynamics inform management decisions in a threatened frog species. *Copeia* 105:53–63.

Le Conte, J. E. 1855. Descriptive catalogue of the Ranina of the United States. *Proceedings of the Academy of Natural Sciences of Philadelphia* 7:423–431.

Lodato, M. J., and D. S. Dugas. 2013. Geographic distribution: *Lithobates areolatus* (Crawfish Frog). *Herpetological Review* 44:103.

Mabee, P. M. 2000. The usefulness of ontogeny in interpreting morphological characters. In *Phylogenetic analysis of morphological data*, ed. J. J. Wiens, 84–114. Washington, DC: Smithsonian Institution Press.

MacArthur, R., and E. O. Wilson. 1967. *Island biogeography theory*. Princeton, NJ: Princeton University Press.

Macey, J. R., J. A. Schulte II, J. L. Strasburg, J. A. Brisson, A. Larson, N. B. Ananjeva, Y. Wang, J. F. Parham, and T. J. Pappenfuss. 2006. Assembly of the eastern North American herpetofauna: new evidence from lizards and frogs. *Biological Letters* 2:388–392.

Maclean, N. 1992. *Young men and fire*. Chicago, IL: University of Chicago Press.

Maglia, A. M., L. A. Pugener, and L. Trueb. 2001. Comparative development of anurans: using phylogeny to understand ontogeny. *American Zoologist* 41:538–551.

McCarty, J. P. 2001. Ecological consequences of recent climate change. *Conservation Biology* 15:320–331.

Mills, L. S. 2007. *Conservation of wildlife populations: demography, genetics, and management*. Malden, MA: Wiley-Blackwell Publishing.

Minton, Jr., S. A. 1972. *Amphibians and reptiles of Indiana*. Indiana Academy of Science, Indiana Academy of Science Monograph, No. 3, Indianapolis, IN.

Minton, Jr., S. A. 2001. *Amphibians and reptiles of Indiana*, 2nd ed. Indianapolis, IN: Indiana Academy of Science.

Minton, Jr., S. A., J. C. List, and M. J. Lodato. 1982. Recent records and status of amphibians and reptiles in Indiana. *Proceedings of the Indiana Academy of Science* 92:489–498.

Mittleman, M. B. 1947. Miscellaneous notes on Indiana amphibians and reptiles. *American Midland Naturalist* 38:466–484.

Morey, S., and D. Reznick. 2001. Effects of larval density on postmetamorphic Spadefoot Toads (*Spea hammondii*). *Ecology* 82:510–522.

Muths, E., T. Chambert, B. R. Schmidt, D. A. Miller, B. R. Hossack, P. Joly, O. Grolet, D. M. Green, D. S. Pilliod, M. Cheylan, R. N. Fisher, R. M. McCaffery, M. J. Adams, W. J. Palen, J. W. Arntzen, J. Garwood, G. Fellers, J. M. Thiron, A. Besnard, and E. H. Campbell Grant. 2017. Heterogeneous responses of temperate-zone amphibian populations to climate change complicates conservation planning. *Scientific Reports*. doi:10.1038/s41598-017-17105-7.

Neill, W. T. 1957. The status of *Rana capito stertens* Schwartz and Harrison. *Herpetologica* 13:47–52.

Nelson, R. K. 1992. *The island within*. Visalia, CA: Vintage Press.

Nomura, F., D. D. Rossa-Feres, and F. Langeani. 2009. Burrowing behavior of *Dermatonotus muelleri* (Anura, Microhylidae) with reference to the origin of burrowing behavior of Anura. *Journal of Ethology* 27:195–201.

Norrocky, M. J. 1984. Burrowing crayfish trap. *Ohio Journal of Science* 84:65–66.

Noss, R. 1996. The naturalists are dying off. *Conservation Biology* 10:1–3.

Nunziata, S. O., M. J. Lannoo, J. R. Robb, D. R. Karns, S. L. Lance, and S. C. Richter. 2013. Population and conservation genetics of Crawfish Frogs, *Lithobates areolatus*, at their northeastern range limit. *Journal of Herpetology* 47:361–368.

O'Connor, M., D. R. Norris, G. T. Crossin, and S. J. Cooke. 2014. Biological carryover effects: linking common concepts and mechanisms in ecology and evolution. *Ecosphere* 5:1–11.

Ordonez, A., J. W. Williams, and J.-C. Svenning. 2016. Mapping climate mechanisms likely to favour the emergence of novel communities. *Nature Climate Change* 6:1104–1109.

Parmesan, C. 2006. Ecological and evolutionary responses to recent climate change. *Annual Review of Ecology, Evolution, and Systematics* 17:637–669.

Parris, M. J. 1998. Terrestrial burrowing ecology of newly metamorphosed frogs (*Rana pipiens* complex). *Canadian Journal of Zoology* 76:2124–2129.

Parris, M. J., and M. Redmer. 2005. *Rana areolata*, Crawfish Frog. In *Amphibian declines: the conservation of United States species*, ed. M. J. Lannoo, 256–258. Berkeley, CA: University of California Press.

Pasukonis, A., M.-C. Loretto, and W. Hödl. 2018. Map-like navigation from distances exceeding routine movements in the Three-striped Poison Frog (*Ameerega trivittata*). *Journal of Experimental Biology*. doi:10.1242/jeb.169714.

Pauley, G. B., D. M. Hillis, and D. C. Cannatella. 2009. Taxonomic freedom and the role of official lists of species names. *Herpetologica* 65:115–128.

Pechenik, J. A., D. E. Wendt, and J. N. Jarrett. 1998. Metamorphosis is not a new beginning: larval experience influences juvenile performance. *BioScience* 48:901–910.

Pechmann, J. H. K. 1994. Population regulation in complex life cycles: aquatic and terrestrial density-dependence in pond-breeding amphibians. Ph.D. dissertation, Duke University.

Pechmann, J. H. K., D. E. Scott, R. D. Semlitsch, J. P. Caldwell, L. J. Vitt, and J. W. Gibbons. 1991. Declining amphibian populations: the problem of separating human impacts from natural fluctuations. *Science* 253:892–895.

Peplow, M. 2004. Sunspot record reveals sun's past: solar history may have links with Earth's climate. *Nature*. doi:10.1038/news041025-15.

Pessier, A. P., and J. R. Mendelson III, eds. 2017. *A manual for control of infectious diseases in amphibian survival assurance colonies and reintroduction programs*, Ver. 2.0. Apple Valley, MN: IUCN/SSC Conservation Breeding Specialist Group. Updated March 2017 by A. P. Pessier, J. R. Mendelson, B. Tapley, and M. Goetz. http://www.cbsg.org (accessed 5 October, 2018).

Phillips, J. B., S. C. Borland, M. J. Freake, J. Brassart, and J. L. Kirschvink. 2002. 'Fixed-axis' magnetic orientation by an amphibian: non-shoreward-directed compass orientation, misdirected homing or positioning a magnetic-based map detector in a consistent alignment relative to the magnetic field? *Journal of Experimental Zoology* 205:3903–3914.

Pillai, P., A. Gonzalez, and M. Loreau. 2012. Evolution of dispersal in a predator-prey metacommunity. *American Naturalist* 179:204–216.

Powell, R., R. Conant, and J. T. Collins. 2016. *Peterson field guide to reptiles and amphibians of eastern and central North America*, 4th ed. Boston, MA: Houghton Mifflin Harcourt.

Prine, J. 1971. Paradise. Atlantic Records, New York, NY.

Pyron, A. R., and J. J. Wiens. 2011. A large-scale phylogeny of Amphibia including over 2800 species, and a revised classification of extant frogs, salamanders, and caecilians. *Molecular Phylogenetics and Evolution* 61:543–583.

Quammen, D. 2013. The short happy life of a Serengeti lion. *National Geographic Magazine*. http://ngm.nationalgeographic.com/2013/08/serengeti-lions/quammen-text.

Reading, C. J. 1998. The effect of winter temperature on the timing of breeding activity in the Common Toad, *Bufo bufo*. *Oecologia* 117:469–475.

Redmer, M. 1999. Relationships of demography and reproductive characteristics of three species of *Rana* (Amphibia: Anura: Ranidae) in southern Illinois. MS thesis, Southern Illinois University.

Redmer, M. 2000. Demographic and reproductive characteristics of a southern Illinois population of the Crayfish Frog, *Rana areolata*. *Journal of the Iowa Academy of Science* 107:128–133.

Resetar, D. R. R., and A. R. Resetar. 2015. Doctor Elias Francis Shipman and the Hoosier Frog. *Proceedings of the Indiana Academy of Science* 124:89–105.

Rice, F. L., and N. S. Davis. 1878. *Rana circulosa*. In *Manual of the vertebrates of the northern United States*, ed. D. S. Jordan, 355. Chicago, IL: Jansen and McClurg.

Rubin, D. 1965. Amphibians and reptiles of Vigo County, Indiana. MA thesis, Indiana State University.

Schwartz, A., and J. R. Harrison, III. 1956. A new subspecies of Gopher Frog (*Rana capito* Le Conte). *Proceedings of the Biological Society of Washington* 69:135–144.

Semlitsch, R. D., and J. P. Caldwell. 1982. Effects of density on growth, metamorphosis, and survivorship in tadpoles of *Scaphiopus holbrookii*. *Ecology* 63:905–911.

Semlitsch, R. D., D. E. Scott, and J. H. K. Pechmann. 1988. Time and size at metamorphosis related to adult fitness in *Ambystoma talpoideum*. *Ecology* 71:184–192.

Simon, T. P. 2001. Checklist of the crayfish and freshwater shrimp (Decapoda) of Indiana. *Proceedings of the Indiana Academy of Science* 110:104–110.

Sinsch, U., and C. Kirst. 2016. Homeward orientation of displaced newts (*Triturus cristatus, Lissotriton vulgaris*) is restricted to the range of routine movements. *Ethology, Ecology, and Evolution* 28:312–328.

Smith, D. C. 1983. Factors controlling tadpole populations of the Chorus Frog (*Pseudacris triseriata*) on Isle Royale, Michigan. *Ecology* 64:501–510.

Smith, H. M. 1950. Handbook of amphibians and reptiles of Kansas. University of Kansas Museum of Natural History, Miscellaneous Publication, No. 2, Lawrence.

Smith, H. M., C. W. Nixon, and P. E. Smith. 1948. A partial description of the tadpole of *Rana areolata circulosa* and notes on the natural history of the race. *American Midland Naturalist* 39:608–614.

Smith, P. W. 1961. The amphibians and reptiles of Illinois. *Illinois Natural History Survey Bulletin* 28(1). Urbana.

Smith, P. W., and S. A. Minton, Jr. 1957. A distributional summary of the herpetofauna of Indiana and Illinois. *American Midland Naturalist* 58:341–351.

Solanki, S. K., I. G. Usoskin, B. Kromer, M. Schüssler, and J. Beer. 2004. Unusual activity of the sun during recent decades compared to the previous 11,000 years. *Nature* 431:1084–1087.

Springer, L. E. 2016. Home range and activity patterns in the Southern Crawfish Frog (*Rana areolata*). MS thesis, Stephen F. Austin State University.

Sredl, M. J., and J. P. Collins. 1992. The interaction of predation, competition, and habitat complexity in structuring an amphibian community. *Copeia* 1992:607–614.

Stiles, R. M., T. R. Halliday, N. J. Engbrecht, J. W. Swan, and M. J. Lannoo. 2017. Wildlife cameras reveal high resolution activity patterns in threatened Crawfish Frogs (*Lithobates areolatus*). *Herpetological Conservation Biology* 12:160–170.

Stiles, R. M., and M. J. Lannoo. 2015. *Lithobates sphenocephalus* (Southern Leopard Frog): fall breeding. *Herpetological Review* 46:414.

Stiles, R. M., M. Sieggreen, A. Preston, A. P. Pessier, S. J. Lannoo, and M. J. Lannoo. 2016b. First report of ranavirus-associated mortality in Crawfish Frogs (*Lithobates areolatus*), a species of conservation concern, in Indiana, USA. *Herpetological Review* 47:389–391.

Stiles, R. M., M. J. Sieggreen, R. A. Johnson, K. Pratt, M. Vassallo, M. Andrus, M. Perry, J. W. Swan, and M. J. Lannoo. 2016c. Captive-rearing state endangered Crawfish Frogs *Lithobates areolatus* from Indiana, USA. *Conservation Evidence* 13:7–11.

Stiles, R. M, J. W. Swan, J. L. Klemish, and M. J. Lannoo. 2016a. Amphibian habitat creation on post-industrial landscapes: a case study in a reclaimed strip-mine area. *Canadian Journal of Zoology*. doi:10.1139/cjz-2015-0163.

Stiles, R. M., V. C. K. Terrell, J. C. Maerz, and M. J. Lannoo. Survivorship, fitness, and carry-over effects in Crawfish Frogs (*Rana areolata*), a species of conservation concern. *Copeia* (in press).

Stuart, S. N., J. S. Chanson, J. A. Cox, B. E. Young, A. S. L. Rodrigues, D. L. Fischman, and R. W. Waller. 2004. Status and trends of amphibian declines and extinctions worldwide. *Science* 306:1783–1786.

Swanson, P. L. 1939. Herpetological notes from Indiana. *American Midland Naturalist* 22:684–695.

Terrell, V. C. K., N. J. Engbrecht, A. P. Pessier, and M. J. Lannoo. 2014a. Drought reduces chytrid fungus (*Batrachochytridium dendrobatidis*) infection intensity and mortality but not prevalence in adult Crawfish Frogs (*Lithobates areolatus*). *Journal of Wildlife Diseases* 50:56–62.

Terrell, V. C. K., J. L. Klemish, N. J. Engbrecht, J. A. May, P. J. Lannoo, R. M. Stiles, and M. J. Lannoo. 2014b. Amphibian and reptile colonization of reclaimed coal spoil grasslands. *Journal of North American Herpetology* 2014:59–68.

Thoma, R. F., and B. J. Armitage. 2008. Burrowing crayfish of Indiana. Final report. Indiana Department of Natural Resources, Indianapolis, IN.

Thompson, C. 1915. Notes on the habits of *Rana areolata* Baird and Girard. Occasional Papers of the Museum of Zoology, No. 10. University of Michigan, Ann Arbor.

Timm, A. 2001. Frog and toad populations in reclaimed and unreclaimed areas of southwestern Indiana, Sullivan and Pike counties. Unpublished report. Indiana University, Bloomington, IN.

Transeau, E. N. 1935. The prairie peninsula. *Ecology* 16:423–437.

Trauth, S. E., H. W. Robison, and M. V. Plummer. 2004. *The amphibians and reptiles of Arkansas*. Fayetteville, NC: University of Arkansas Press.

Urban, M. C., J. L. Richardson, and N. A. Freidenfelds. 2013. Plasticity and genetic adaptation mediate amphibian and reptile response to climate change. *Evolutionary Applications* 7:88–103.

Van Allen, B. G., V. S. Briggs, M. W. McCoy, and J. R. Vonesh. 2010. Carry-over effects of the larval environment on post-metamorphic performance in two hylid frogs. *Oecologia* 164:891–898.

Walls, S. C., W. J. Barichivich, and M. E. Brown. 2013. Drought, deluge and declines: the impact of precipitation extremes on amphibians in a changing climate. *Biology* 2:399–418.

Welch, S. M., and A. G. Eversole. 2006. Comparison of two burrowing crayfish trapping methods. *Southeastern Naturalist* 5:27–30.

Wells, K. D. 2007. *The ecology and behavior of amphibians*. Chicago, IL: University of Chicago Press.

Werner, E. E. 1992. Competitive interactions between Wood Frog and Northern Leopard Frog larvae. *Ecology* 77:157–169.

Whittaker, K., and V. Vredenburg. 2011. An overview of Chytridiomycosis: update: *Batrachochytridium salamandrivorans*: deadly fungal threat to amphibians. https://amphibiaweb.org/chytrid/chytridiomycosis.html (accessed 16 May, 2019).

Wilbur, H. M. 1972. Competition, predation, and the structure of the *Ambystoma-Rana sylvatica* community. *Ecology* 53:3–21.

Wilbur, H. M. 1976. Density-dependent aspects of metamorphosis in *Ambystoma* and *Rana sylvatica*. *Ecology* 57:1289–1296.

Wilbur, H. M. 1977a. Density-dependent aspects of growth and metamorphosis in *Bufo americanus*. *Ecology* 58:196–200.

Wilbur, H. M. 1977b. Interactions of food level and population density in *Rana sylvatica*. *Ecology* 58:206–209.

Wilbur, H. M. 1982. Competition between tadpoles of *Hyla femoralis* and *Hyla gratiosa* in laboratory experiments. *Ecology* 63:278–282.

Wilbur, H. M. 1997. Experimental ecology of food webs: complex systems in temporary ponds. *Ecology* 78:2279–2302.

Wilbur, H. M., and R. A. Alford. 1985. Priority effects in experimental pond communities: responses of *Hyla* to *Bufo* and *Rana*. *Ecology* 66:1106–1114.

Wilbur, H. M., and J. P. Collins. 1973. Ecological aspects of amphibian metamorphosis. *Science* 182:1305–1314.

Wilbur, H. M., P. J. Morin, and R. N. Harris. 1983. Salamander predation and the structure of experimental communities: anuran responses. *Ecology* 64:1423–1429.

Williams, P. J., N. J. Engbrecht, J. R. Robb, V. C. K. Terrell, and M. J. Lannoo. 2013. Surveying a threatened amphibian species through a narrow detection window. *Copeia* 2013:552–561.

Williams, P. J., J. R. Robb, R. H. Kappler, T. E. Piening, and D. R. Karns. 2012. Intraspecific density dependence in larval development of the Crawfish Frog, *Lithobates areolatus*. *Herpetological Review* 43:36–38.

Woodin, S. A., T. J. Hilbish, B. Helmuth, S. J. Jones, and D. S. Wethey. 2013. Climate change, species distribution models, and physiological performance metrics: predicting when biogeographic models are likely to fail. *Ecology and Evolution* 3:3334–3346.

Wright, A. H. 1932. *Life histories of the frogs of the Okefinokee Swamp, Georgia*. North American Salientia (Anura) No. 2. New York, NY: Macmillan Press.

Wright, A. H., and A. A. Wright. 1933. *Handbook of frogs and toads of the United States and Canada*. Ithaca, NY: Comstock Publishing Associates.

Wright, A. H., and A. A. Wright. 1942. *Handbook of frogs and toads: the frogs and toads of the United States and Canada*. Ithaca, NY: Comstock Press.

Wright, A. H., and A. A. Wright. 1949. *Handbook of frogs and toads: the frogs and toads of the United States and Canada*. Ithaca, NY: Comstock Press.

Wright, H. P., and G. S. Myers. 1927. *Rana areolata* at Bloomington, Indiana. *Copeia* 159:173–175.

Yagi, K. T., and D. M. Green. 2018. Post-metamorphic carry-over effects in a complex life history: behavior and growth and two life stages in an amphibian, *Anaxyrus fowleri*. *Copeia* 106:77–85.

Young, J., B. I. Crother, and J. D. McEachran. 2001. Allozyme evidence for the separation of *Rana areolata* and *Rana capito* and for the resurrection of *Rana sevosa*. *Copeia* 2001:382–388.

Zaret, T. M. 1980. *Predation and freshwater communities*. New Haven, CT: Yale University Press.

Index

Printed in the United States
by Baker & Taylor Publisher Services